Fifty Year Almanac
of
Astronomical Events
2021 To 2070

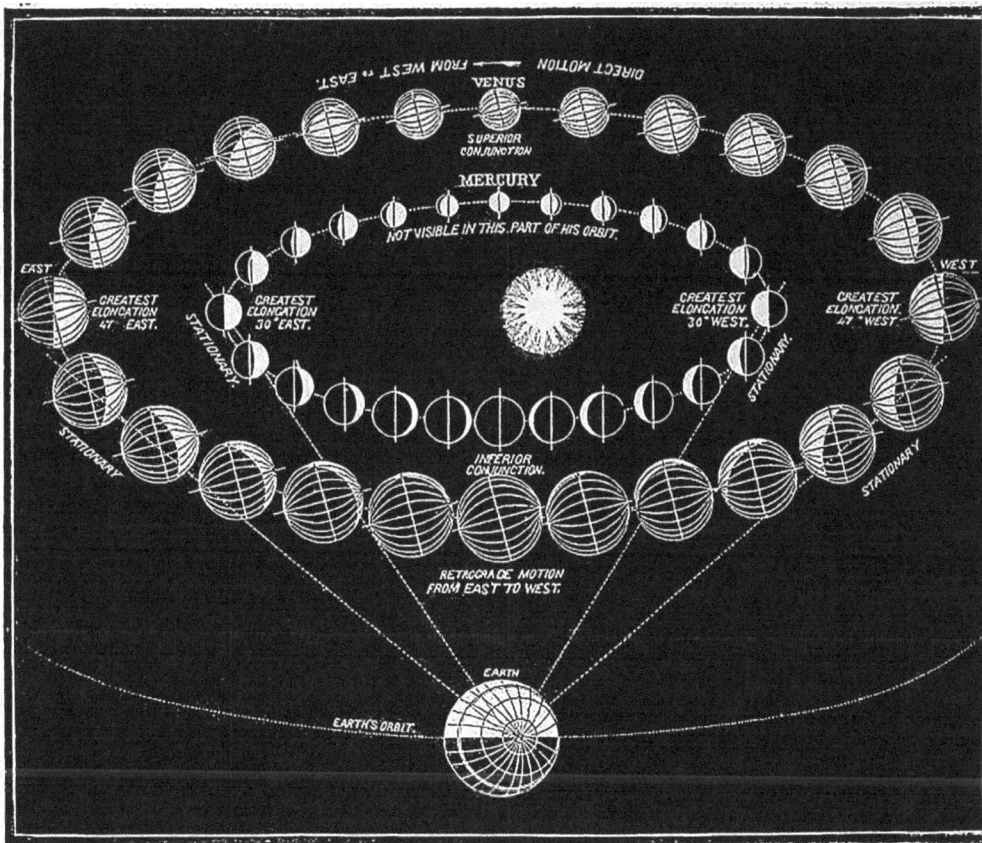

Fred Espenak

Edition 1.0
January 2021

Fifty Year Almanac of Astronomical Events – 2021 To 2070

Astropixels Publishing
P.O. Box 16197
Portal, AZ 85632

www.astropixels.com/pubs

For more about this book see: www.astropixels.com/pubs/FiftyYearAlmanac.html

More about astronomical events from 2001 to 2100 can be found at astropixels.com:

www.astropixels.com/almanac/almanac.html

Astropixels Publication Number: AP035

First Edition (Version 1.0a)

ISBN   978-1-941983-35-5

Printed in the United States of America

Front Cover (foreground): Inner Planets, from *Smith's Illustrated Astronomy*, Asa Smith, 1885.
Front Cover (background): Table of Astronomy, from *Cyclopaedia,* Ephraim Chambers, 1728.

Back Cover: Portrait of Fred Espenak (Copyright ©2018 by Fred Espenak).

Introduction

The *Fifty Year Almanac of Astronomical Events* presents a wide range of solar system phenomena as seen from Earth. Each year gives a concise list of the most conspicuous and/or significant astronomical events involving the Sun, Moon and planets.

These astronomical events include:

> - solar and lunar eclipses
> - phases of the Moon
> - apogees and perigees of the Moon
> - Equinoxes and Solstices of Earth
> - aphelion and perihelion (Earth, Mars, Jupiter, Saturn and Uranus)
> - oppositions and conjunctions of the planets
> - elongations of Mercury and Venus
> - close conjunctions of the Moon with the planets and bright stars
> - close conjunctions of planets with bright stars and other planets
> - the peak of major meteor showers

The *Fifty Year Almanac of Astronomical Events* is based on the web resource *Sky Events Almanacs* on AstroPixels.com (*www.astropixels.com/almanac/almanac.html*). This website contains Sky Event Almanacs for the entire 21st Century, and for 24 time zones around the globe.

Algorithms used in predicting many of the astronomical events are based on *Astronomical Algorithms* by Jean Meeus (Willmann-Bell Inc. Richmond 1998). In some cases where higher accuracy is required, the JPL DE405 is used in calculating planetary positions.

The goal of the *Sky Event Almanacs* is to present a wide range of solar system phenomena with practicable accuracy. In general, events listed to the nearest hour are accurate to ± 30 minutes. Events listed with a precision in hours and minutes (i.e., hh:mm) are typically accurate to ± 5 minutes or better.

Glossary of Terms Used in Fifty Year Almanac

Perihelion - instant in a planet's orbit when a planet is closest to the Sun
Aphelion - instant in a planet's orbit when a planet is furthest from the Sun
Perigee - instant in a Moon's orbit when the Moon is closest to Earth
Apogee - instant in a Moon's orbit when the Moon is furthest from Earth
Inferior Conjunction - instant when a planet (Mercury or Venus) passes between Earth and the Sun
Superior Conjunction - instant when a planet (Mercury or Venus) passes on the opposite side of the Sun from Earth
Greatest Elongation - the maximum angular separation between the Sun and the planet (Mercury or Venus) as seen from Earth
— during eastern elongation (E), the planet appears as an evening star
— during western elongation (W), the planet appears as a morning star
Opposition - instant when a planet appears opposite the Sun as seen from Earth
Conjunction - instant when a planet appears closest to the Sun as seen from Earth
Occultation (Occn.) - the Moon occults or eclipses a star or planet

Aldebaran - bright star in the constellation Taurus
Pollux - bright star in the constellation Gemini
Regulus - bright star in the constellation Leo
Spica - bright star in the constellation Virgo
Antares - bright star in the constellation Scorpius
Pleiades - bright star cluster in the constellation Taurus

Greenwich Mean Time and Time Zones

The date and time of each event in the *Fifty Year Almanac* is given in Greenwich Mean Time – the time measured on Earth's zero degree line of longitude, or meridian. Greenwich Mean Time (GMT) is set equal to Coordinated Universal Time (UTC) – a time system based on atomic time and synchronized with the rotation of Earth.

The sky events listed in the Almanac can be converted to any other time zone by noting the time zone's offset from Coordinated Universal Time (UTC). Because Greenwich Mean Time is set equal to UTC, its offset is 0.0 hours. Below is a sampling of common time zones in use around the world along with their offsets in hours from UTC.

Table 1 gives a list of commonly used time zones for the Eastern Hemisphere.

Table 1– Eastern Hemisphere: Time Zones, Abbreviations, and UTC Offsets

Time Zone Name	Abbrev.	UTC Offset
Greenwich Mean Time	GMT	UTC + 0 hr
Central European Time	CET	UTC + 1 hr
Eastern European Time	EET	UTC + 2 hrs
Moscow Time	MSK	UTC + 3 hrs
Gulf Standard Time	GST	UTC + 4 hrs
Pakistan Standard Time	PKT	UTC + 5 hrs
Indian Standard Time	IST	UTC + 5.5 hrs
Bangladesh Standard Time	BST	UTC + 6 hrs
Indochina Time	ICT	UTC + 7 hrs
China Standard Time	CST	UTC + 8 hrs
Australian Western Standard Time	AWST	UTC + 8 hrs
Japan Standard Time	JST	UTC + 9 hrs
Australian Central Time	ACT	UTC + 9.5 hrs
Australian Eastern Standard Time	AEST	UTC + 10 hrs
New Caledonia Time	NCT	UTC + 11 hrs
New Zealand Standard Time	NZST	UTC + 12 hrs

Table 2 gives a list of commonly used time zones for the Western Hemisphere.

Table 2 – Western Hemisphere: Time Zones, Abbreviations, and UTC Offsets

Time Zone Name	Abbrev.	UTC Offset
Cape Verde Time	CVT	UTC - 1 hr
South Georgia Time	GST	UTC - 2 hrs
Argentina Time	ART	UTC - 3 hrs
Atlantic Standard Time	AST	UTC - 4 hrs
Eastern Standard Time	EST	UTC - 5 hrs
Central Standard Time	CST	UTC - 6 hrs
Mountain Standard Time	MST	UTC - 7 hrs
Pacific Standard Time	PST	UTC - 8 hrs
Alaskan Standard Time	AKST	UTC - 9 hrs
Hawaiian Standard Time	HST	UTC - 10 hrs
Samoa Standard Time	SST	UTC - 11 hrs
International Day Line West	IDLW	UTC - 12 hrs

For a more extensive list of time zones used around the world, see: *www.timeanddate.com/time/zones/*

Almanac of Astronomical Events for 2021

Date	GMT (h:m)	Event
Jan 02	14	Earth at Perihelion: 0.98325 AU
03	15	Quadrantid Meteor Shower
06	09:37	LAST QUARTER MOON
09	15:39	Moon at Perigee: 367390 km
11	20:12	Venus 1.5°N of Moon
13	05:00	NEW MOON
14	08:15	Mercury 2.3°N of Moon
20	21:02	FIRST QUARTER MOON
21	13:11	Moon at Apogee: 404361 km
24	02	Mercury at Greatest Elong: 18.6°E
24	02	Saturn in Conjunction with Sun
27	15:46	Pollux 3.8°N of Moon
28	19:16	FULL MOON
29	01	Jupiter in Conjunction with Sun
Feb 03	19:33	Moon at Perigee: 370127 km
04	17:37	LAST QUARTER MOON
08	14	Mercury at Inferior Conjunction
10	11:16	Saturn 3.4°N of Moon
11	19:06	NEW MOON
18	10:22	Moon at Apogee: 404467 km
18	22:47	Mars 3.7°N of Moon
19	18:47	FIRST QUARTER MOON
23	08	Mercury 4.0°N of Saturn
24	01:10	Pollux 3.7°N of Moon
27	08:17	FULL MOON
Mar 02	05:19	Moon at Perigee: 365422 km
03	23:36	Mars 2.6°S of Pleiades
05	05	Mercury 0.3°N of Jupiter
06	01:30	LAST QUARTER MOON
06	11	Mercury at Greatest Elong: 27.3°W
09	23:02	Saturn 3.7°N of Moon
10	15:35	Jupiter 4.0°N of Moon
11	00	Neptune in Conjunction with Sun
11	01:02	Mercury 3.7°N of Moon
13	10:21	NEW MOON
18	05:04	Moon at Apogee: 405253 km
19	17:48	Mars 1.9°N of Moon
20	09:37	Vernal Equinox
21	14:40	FIRST QUARTER MOON
23	10:26	Pollux 3.5°N of Moon
26	06	Venus at Superior Conjunction
28	18:48	FULL MOON
30	06:12	Moon at Perigee: 360311 km

Date	GMT (h:m)	Event
Apr 04	10:02	LAST QUARTER MOON
06	08:34	Saturn 4.0°N of Moon
07	07:15	Jupiter 4.4°N of Moon
12	02:31	NEW MOON
14	17:47	Moon at Apogee: 406120 km
17	12:09	Mars 0.1°N of Moon: Occn.
19	02	Mercury at Superior Conjunction
19	18:18	Pollux 3.2°N of Moon
20	06:59	FIRST QUARTER MOON
22	12	Lyrid Meteor Shower
27	03:31	FULL MOON
27	15:24	Moon at Perigee: 357379 km
30	21	Uranus in Conjunction with Sun
May 03	17:02	Saturn 4.2°N of Moon
03	19:50	LAST QUARTER MOON
04	03:03	Mercury 2.1°S of Pleiades
04	21:00	Jupiter 4.6°N of Moon
05	01	Eta-Aquarid Meteor Shower
11	19:00	NEW MOON
11	21:54	Moon at Apogee: 406512 km
13	17:59	Mercury 2.1°N of Moon
16	04:47	Mars 1.5°S of Moon
17	00:39	Pollux 3.1°N of Moon
17	06	Mercury at Greatest Elong: 22.0°E
19	19:13	FIRST QUARTER MOON
26	01:52	Moon at Perigee: 357310 km
26	11:14	FULL MOON
26	11:19	Total Lunar Eclipse; mag=1.009
29	03	Mercury 0.4°S of Venus
31	01:22	Saturn 4.2°N of Moon
Jun 01	08:57	Jupiter 4.6°N of Moon
02	07:24	LAST QUARTER MOON
08	02:27	Moon at Apogee: 406230 km
10	10:42	Annular Solar Eclipse; mag=0.943
10	10:53	NEW MOON
11	01	Mercury at Inferior Conjunction
12	06:44	Venus 1.5°S of Moon
13	06:19	Pollux 3.1°N of Moon
13	19:52	Mars 2.8°S of Moon
18	03:54	FIRST QUARTER MOON
21	03:32	Summer Solstice
23	09:58	Moon at Perigee: 359960 km
24	18:40	FULL MOON
27	09:30	Saturn 4.0°N of Moon
28	18:38	Jupiter 4.5°N of Moon

Date	GMT (h:m)	Event
Jul 01	21:11	LAST QUARTER MOON
04	20	Mercury at Greatest Elong: 21.6°W
05	14:48	Moon at Apogee: 405342 km
05	23	Earth at Aphelion: 1.01673 AU
08	04:38	Mercury 3.7°S of Moon
10	01:17	NEW MOON
12	09:10	Venus 3.3°S of Moon
12	10:10	Mars 3.8°S of Moon
13	00	Mars at Aphelion: 1.66596 AU
13	13	Venus 0.5°N of Mars
17	10:11	FIRST QUARTER MOON
21	10:30	Moon at Perigee: 364520 km
21	21:21	Venus 1.0°N of Regulus
24	02:37	FULL MOON
24	16:42	Saturn 3.8°N of Moon
26	01:17	Jupiter 4.2°N of Moon
28	03	Delta-Aquarid Meteor Shower
29	14:09	Mars 0.6°N of Regulus
31	13:16	LAST QUARTER MOON
Aug 01	14	Mercury at Superior Conjunction
02	05	Saturn at Opposition
02	07:35	Moon at Apogee: 404412 km
06	19:42	Pollux 3.1°N of Moon
08	13:50	NEW MOON
10	00:42	Mars 4.3°S of Moon
11	07:00	Venus 4.3°S of Moon
12	19	Perseid Meteor Shower
15	15:20	FIRST QUARTER MOON
17	09:23	Moon at Perigee: 369127 km
19	03	Mercury 0.1°S of Mars
19	23	Jupiter at Opposition
20	22:19	Saturn 3.7°N of Moon
22	04:52	Jupiter 4.0°N of Moon
22	12:02	FULL MOON
30	02:22	Moon at Apogee: 404100 km
30	07:13	LAST QUARTER MOON
Sep 03	04:04	Pollux 3.0°N of Moon
05	14:32	Venus 1.4°N of Spica
07	00:52	NEW MOON
10	02:09	Venus 4.1°S of Moon
11	10:05	Moon at Perigee: 368464 km
12	23:59	Antares 4.2°S of Moon
13	20:39	FIRST QUARTER MOON
14	04	Mercury at Greatest Elong: 26.8°E
14	08	Neptune at Opposition
17	02:37	Saturn 3.8°N of Moon
18	06:50	Jupiter 4.0°N of Moon
20	23:55	FULL MOON
21	02:00	Mercury 1.2°S of Spica
22	19:21	Autumnal Equinox
26	21:44	Moon at Apogee: 404641 km
29	01:57	LAST QUARTER MOON
30	12:44	Pollux 2.8°N of Moon

Date	GMT (h:m)	Event
Oct 06	11:05	NEW MOON
08	04	Mars in Conjunction with Sun
08	17:28	Moon at Perigee: 363388 km
09	16	Mercury at Inferior Conjunction
09	18:36	Venus 2.9°S of Moon
10	06:27	Antares 4.0°S of Moon
13	03:25	FIRST QUARTER MOON
14	07:12	Saturn 3.9°N of Moon
15	09:58	Jupiter 4.1°N of Moon
16	13:24	Venus 1.4°N of Antares
20	14:57	FULL MOON
21	11	Orionid Meteor Shower
24	15:30	Moon at Apogee: 405616 km
25	05	Mercury at Greatest Elong: 18.4°W
27	20:40	Pollux 2.6°N of Moon
28	20:05	LAST QUARTER MOON
29	22	Venus at Greatest Elong: 47.0°E
Nov 03	18:40	Mercury 1.2°S of Moon: Occn.
04	21:15	NEW MOON
05	00	Uranus at Opposition
05	12	S Taurid Meteor Shower
05	22:23	Moon at Perigee: 358845 km
06	15:29	Antares 3.9°S of Moon
08	05:21	Venus 1.1°S of Moon: Occn.
10	14:27	Saturn 4.1°N of Moon
11	12:46	FIRST QUARTER MOON
11	17:13	Jupiter 4.4°N of Moon
12	11	N Taurid Meteor Shower
17	18	Leonid Meteor Shower
19	08:58	FULL MOON
19	09:03	Partial Lunar Eclipse; mag=0.974
21	02:14	Moon at Apogee: 406276 km
24	03:22	Pollux 2.5°N of Moon
27	12:28	LAST QUARTER MOON
29	05	Mercury at Superior Conjunction
Dec 03	00:28	Mars 0.7°S of Moon: Occn.
04	07:33	Total Solar Eclipse; mag=1.037
04	07:43	NEW MOON
04	10:01	Moon at Perigee: 356794 km
07	00:48	Venus 1.9°N of Moon
08	01:52	Saturn 4.2°N of Moon
09	06:07	Jupiter 4.5°N of Moon
11	01:36	FIRST QUARTER MOON
14	07	Geminid Meteor Shower
18	02:16	Moon at Apogee: 406322 km
19	04:36	FULL MOON
21	09:20	Pollux 2.6°N of Moon
21	15:59	Winter Solstice
22	15	Ursid Meteor Shower
27	02:24	LAST QUARTER MOON
29	05	Mercury 4.2°S of Venus
31	13:54	Antares 3.9°S of Moon
31	20:13	Mars 1.0°N of Moon: Occn.

Almanac of Astronomical Events for 2022

Date	GMT (h:m)	Event
Jan 01	23:00	Moon at Perigee: 358037 km
02	18:33	NEW MOON
03	21	Quadrantid Meteor Shower
04	01:23	Mercury 3.1°N of Moon
04	07	Earth at Perihelion: 0.98333 AU
04	16:50	Saturn 4.2°N of Moon
06	00:09	Jupiter 4.5°N of Moon
07	11	Mercury at Greatest Elong: 19.2°E
09	01	Venus at Inferior Conjunction
09	18:11	FIRST QUARTER MOON
13	04	Mercury 3.4°N of Saturn
14	09:27	Moon at Apogee: 405806 km
17	15:37	Pollux 2.6°N of Moon
17	23:49	FULL MOON
23	10	Mercury at Inferior Conjunction
25	13:41	LAST QUARTER MOON
27	22:57	Antares 3.7°S of Moon
29	15:05	Mars 2.4°N of Moon
30	07:09	Moon at Perigee: 362250 km
Feb 01	05:46	NEW MOON
02	21:08	Jupiter 4.3°N of Moon
04	19	Saturn in Conjunction with Sun
08	13:50	FIRST QUARTER MOON
11	02:39	Moon at Apogee: 404897 km
13	22:52	Pollux 2.6°N of Moon
16	16:57	FULL MOON
16	21	Mercury at Greatest Elong: 26.3°W
23	22:32	LAST QUARTER MOON
24	05:17	Antares 3.5°S of Moon
26	22:18	Moon at Perigee: 367787 km
27	09:00	Mars 3.5°N of Moon
28	20:07	Mercury 3.7°N of Moon
28	23:47	Saturn 4.3°N of Moon
Mar 02	16	Mercury 0.7°S of Saturn
02	17:35	NEW MOON
05	13	Jupiter in Conjunction with Sun
08	16:46	Pleiades 3.8°N of Moon
10	10:45	FIRST QUARTER MOON
10	23:05	Moon at Apogee: 404268 km
13	06:58	Pollux 2.4°N of Moon
16	04	Venus 3.9°N of Mars
18	07:17	FULL MOON
20	10	Venus at Greatest Elong: 46.6°W
20	15:33	Vernal Equinox
23	10:43	Antares 3.2°S of Moon
23	23:28	Moon at Perigee: 369764 km
25	05:37	LAST QUARTER MOON
28	02:54	Mars 4.1°N of Moon
28	11:43	Saturn 4.4°N of Moon
29	01	Venus 2.1°N of Saturn
30	14:34	Jupiter 3.9°N of Moon
Apr 01	06:24	NEW MOON
02	23	Mercury at Superior Conjunction
Apr 05	01:16	Pleiades 3.6°N of Moon
05	02	Mars 0.3°S of Saturn
07	19:11	Moon at Apogee: 404438 km
09	06:47	FIRST QUARTER MOON
09	15:14	Pollux 2.2°N of Moon
16	18:55	FULL MOON
19	15:16	Moon at Perigee: 365143 km
19	17:36	Antares 3.1°S of Moon
22	18	Lyrid Meteor Shower
23	11:56	LAST QUARTER MOON
24	20:56	Saturn 4.5°N of Moon
25	22:06	Mars 3.9°N of Moon
27	01:51	Venus 3.8°N of Moon
27	08:23	Jupiter 3.6°N of Moon
29	08	Mercury at Greatest Elong: 20.6°E
29	19:31	Mercury 1.3°S of Pleiades
30	20	Venus 0.2°S of Jupiter
30	20:28	NEW MOON
30	20:41	Partial Solar Eclipse; mag=0.640
May 02	09:00	Pleiades 3.6°N of Moon
02	14:17	Mercury 1.8°N of Moon
05	08	Eta-Aquarid Meteor Shower
05	09	Uranus in Conjunction with Sun
05	12:46	Moon at Apogee: 405287 km
06	22:56	Pollux 2.1°N of Moon
09	00:21	FIRST QUARTER MOON
16	04:11	Total Lunar Eclipse; mag=1.414
16	04:14	FULL MOON
17	02:48	Antares 3.1°S of Moon
17	15:23	Moon at Perigee: 360298 km
21	19	Mercury at Inferior Conjunction
22	04:43	Saturn 4.5°N of Moon
22	18:43	LAST QUARTER MOON
24	19:24	Mars 2.8°N of Moon
24	23:59	Jupiter 3.3°N of Moon
27	02:52	Venus 0.2°N of Moon: Occn.
29	09	Mars 0.6°S of Jupiter
30	11:30	NEW MOON
Jun 02	01:14	Moon at Apogee: 406191 km
03	05:42	Pollux 2.1°N of Moon
07	14:48	FIRST QUARTER MOON
13	13:26	Antares 3.1°S of Moon
14	11:52	FULL MOON
14	23:21	Moon at Perigee: 357434 km
16	15	Mercury at Greatest Elong: 23.2°W
18	12:22	Saturn 4.3°N of Moon
21	03:11	LAST QUARTER MOON
21	09:14	Summer Solstice
21	13	Mars at Perihelion: 1.38130 AU
21	13:32	Jupiter 2.7°N of Moon
22	18:08	Mercury 2.8°N of Aldebaran
22	18:16	Mars 0.9°N of Moon: Occn.
25	21:27	Pleiades 3.5°N of Moon
26	08:11	Venus 2.7°S of Moon
27	08:19	Mercury 3.9°S of Moon

Date	GMT (h:m)	Event
Jun 29	02:52	NEW MOON
29	06:08	Moon at Apogee: 406581 km
30	11:46	Pollux 2.2°N of Moon
Jul 04	07	Earth at Aphelion: 1.01672 AU
07	02:14	FIRST QUARTER MOON
10	23:50	Antares 3.0°S of Moon
13	09:08	Moon at Perigee: 357264 km
13	18:37	FULL MOON
15	20:16	Saturn 4.0°N of Moon
16	19	Mercury at Superior Conjunction
19	00:55	Jupiter 2.2°N of Moon
20	14:18	LAST QUARTER MOON
21	16:46	Mars 1.1°S of Moon: Occn.
23	03:29	Pleiades 3.4°N of Moon
26	10:22	Moon at Apogee: 406276 km
26	14:12	Venus 4.2°S of Moon
28	09	Delta-Aquarid Meteor Shower
28	17:55	NEW MOON
Aug 04	04:58	Mercury 0.6°N of Regulus
05	11:06	FIRST QUARTER MOON
07	08:29	Antares 2.8°S of Moon
10	17:14	Moon at Perigee: 359830 km
12	01:36	FULL MOON
12	03:55	Saturn 3.9°N of Moon
13	01	Perseid Meteor Shower
14	17	Saturn at Opposition
15	09:37	Jupiter 1.9°N of Moon
19	04:36	LAST QUARTER MOON
19	10:32	Pleiades 3.1°N of Moon
19	12:16	Mars 2.7°S of Moon
22	21:53	Moon at Apogee: 405419 km
24	00:17	Pollux 2.1°N of Moon
25	20:58	Venus 4.3°S of Moon
27	08:17	NEW MOON
27	16	Mercury at Greatest Elong: 27.3°E
Sep 03	14:56	Antares 2.5°S of Moon
03	18:08	FIRST QUARTER MOON
07	18:17	Moon at Perigee: 364491 km
08	10:31	Saturn 3.9°N of Moon
10	09:59	FULL MOON
11	15:11	Jupiter 1.8°N of Moon
15	18:50	Pleiades 2.9°N of Moon
16	21	Neptune at Opposition
17	01:41	Mars 3.6°S of Moon
17	21:52	LAST QUARTER MOON
19	14:44	Moon at Apogee: 404556 km
20	07:40	Pollux 1.9°N of Moon
23	01:04	Autumnal Equinox
23	07	Mercury at Inferior Conjunction
25	21:54	NEW MOON
26	18	Jupiter at Opposition
27	09:22	Spica 4.3°S of Moon
30	20:20	Antares 2.4°S of Moon

Date	GMT (h:m)	Event
Oct 03	00:14	FIRST QUARTER MOON
04	17:01	Moon at Perigee: 369335 km
05	15:51	Saturn 4.1°N of Moon
08	18:06	Jupiter 2.1°N of Moon
08	21	Mercury at Greatest Elong: 18.0°W
09	20:55	FULL MOON
13	03:46	Pleiades 2.7°N of Moon
15	04:28	Mars 3.6°S of Moon
17	10:21	Moon at Apogee: 404330 km
17	15:41	Pollux 1.8°N of Moon
17	17:15	LAST QUARTER MOON
21	18	Orionid Meteor Shower
22	21	Venus at Superior Conjunction
25	10:49	NEW MOON
25	11:00	Partial Solar Eclipse; mag=0.862
28	02:48	Antares 2.3°S of Moon
29	14:48	Moon at Perigee: 368289 km
Nov 01	06:37	FIRST QUARTER MOON
01	21:08	Saturn 4.2°N of Moon
04	20:19	Jupiter 2.4°N of Moon
08	10:59	Total Lunar Eclipse; mag=1.359
08	11:02	FULL MOON
08	16	Mercury at Superior Conjunction
09	09	Uranus at Opposition
09	12:16	Pleiades 2.7°N of Moon
11	13:43	Mars 2.5°S of Moon
13	23:43	Pollux 1.7°N of Moon
14	06:41	Moon at Apogee: 404924 km
16	13:27	LAST QUARTER MOON
18	00	Leonid Meteor Shower
21	03:36	Spica 4.3°S of Moon
23	22:57	NEW MOON
26	01:30	Moon at Perigee: 362826 km
29	04:40	Saturn 4.2°N of Moon
30	14:36	FIRST QUARTER MOON
Dec 02	00:52	Jupiter 2.5°N of Moon
06	19:26	Pleiades 2.7°N of Moon
08	04:08	FULL MOON
08	04:21	Mars 0.5°S of Moon: Occn.
08	04	Mars at Opposition
11	07:06	Pollux 1.8°N of Moon
12	00:30	Moon at Apogee: 405869 km
14	13	Geminid Meteor Shower
16	08:56	LAST QUARTER MOON
21	15	Mercury at Greatest Elong: 20.1°E
21	21:48	Winter Solstice
21	22:43	Antares 2.3°S of Moon
22	21	Ursid Meteor Shower
23	10:17	NEW MOON
24	08:32	Moon at Perigee: 358270 km
24	11:29	Venus 3.5°N of Moon
24	18:31	Mercury 3.8°N of Moon
26	16:11	Saturn 4.0°N of Moon
29	07	Mercury 1.4°N of Venus
29	10:29	Jupiter 2.3°N of Moon
30	01:21	FIRST QUARTER MOON

Almanac of Astronomical Events for 2023

Date	GMT (h:m)	Event
Jan 03	01:24	Pleiades 2.6°N of Moon
03	19:35	Mars 0.5°N of Moon: Occn.
04	03	Quadrantid Meteor Shower
04	16	Earth at Perihelion: 0.98329 AU
06	23:08	FULL MOON
07	13	Mercury at Inferior Conjunction
07	13:40	Pollux 1.9°N of Moon
08	09:19	Moon at Apogee: 406459 km
15	02:10	LAST QUARTER MOON
18	09:32	Antares 2.1°S of Moon
20	12	Jupiter at Perihelion: 4.95101 AU
21	20:53	NEW MOON
21	20:58	Moon at Perigee: 356570 km
22	22	Venus 0.3°S of Saturn
23	07:22	Saturn 3.8°N of Moon
23	08:20	Venus 3.5°N of Moon
26	02:00	Jupiter 1.8°N of Moon
28	15:19	FIRST QUARTER MOON
30	06	Mercury at Greatest Elong: 25.0°W
30	07:21	Pleiades 2.4°N of Moon
31	04:24	Mars 0.1°N of Moon: Occn.
Feb 03	19:47	Pollux 1.9°N of Moon
04	08:55	Moon at Apogee: 406476 km
05	18:29	FULL MOON
11	04:23	Spica 3.6°S of Moon
13	16:01	LAST QUARTER MOON
14	18:09	Antares 1.9°S of Moon
16	16	Saturn in Conjunction with Sun
18	20:53	Mercury 3.6°N of Moon
19	09:06	Moon at Perigee: 358267 km
20	07:06	NEW MOON
22	07:57	Venus 2.1°N of Moon
22	21:58	Jupiter 1.2°N of Moon: Occn.
26	14:42	Pleiades 2.1°N of Moon
27	08:06	FIRST QUARTER MOON
28	04:32	Mars 1.1°S of Moon: Occn.
Mar 02	04	Venus 0.5°N of Jupiter
03	02:10	Pollux 1.7°N of Moon
03	18:01	Moon at Apogee: 405890 km
07	12:40	FULL MOON
10	10:06	Spica 3.4°S of Moon
14	00:21	Antares 1.6°S of Moon
15	02:08	LAST QUARTER MOON
17	11	Mercury at Superior Conjunction
19	15:16	Moon at Perigee: 362698 km
19	15:20	Saturn 3.6°N of Moon
20	21:25	Vernal Equinox
21	17:23	NEW MOON
22	19:54	Jupiter 0.5°N of Moon: Occn.
24	10:28	Venus 0.1°N of Moon: Occn.
25	23:42	Pleiades 1.9°N of Moon
28	13:16	Mars 2.3°S of Moon
29	02:32	FIRST QUARTER MOON
30	09:23	Pollux 1.6°N of Moon
31	11:18	Moon at Apogee: 404921 km

Date	GMT (h:m)	Event
Apr 06	04:35	FULL MOON
06	16:45	Spica 3.3°S of Moon
10	05:50	Antares 1.5°S of Moon
11	04:42	Venus 2.5°S of Pleiades
11	21	Jupiter in Conjunction with Sun
11	22	Mercury at Greatest Elong: 19.5°E
13	09:11	LAST QUARTER MOON
16	02:22	Moon at Perigee: 367967 km
16	03:47	Saturn 3.5°N of Moon
20	04:12	NEW MOON
20	04:17	Hybrid Solar Eclipse; mag=1.013
22	09:14	Pleiades 1.9°N of Moon
23	01	Lyrid Meteor Shower
23	13:03	Venus 1.3°S of Moon
26	02:18	Mars 3.2°S of Moon
26	17:26	Pollux 1.5°N of Moon
27	21:20	FIRST QUARTER MOON
28	06:43	Moon at Apogee: 404300 km
May 01	23	Mercury at Inferior Conjunction
04	00:55	Spica 3.3°S of Moon
05	14	Eta-Aquarid Meteor Shower
05	17:23	Pen. Lunar Eclipse; mag=0.964
05	17:34	FULL MOON
07	12:35	Antares 1.5°S of Moon
09	21	Uranus in Conjunction with Sun
11	04:57	Moon at Perigee: 369345 km
12	14:28	LAST QUARTER MOON
13	13:04	Saturn 3.3°N of Moon
17	13:15	Jupiter 0.8°S of Moon: Occn.
18	01:34	Mercury 3.6°S of Moon
19	15:53	NEW MOON
23	12:08	Venus 2.2°S of Moon
24	01:37	Pollux 1.6°N of Moon
24	17:32	Mars 3.8°S of Moon
26	01:39	Moon at Apogee: 404510 km
27	15:22	FIRST QUARTER MOON
29	05	Mercury at Greatest Elong: 24.9°W
29	13:04	Venus 3.9°S of Pollux
30	21	Mars at Aphelion: 1.66594 AU
31	10:05	Spica 3.3°S of Moon
Jun 03	21:19	Antares 1.6°S of Moon
04	03:42	FULL MOON
04	11	Venus at Greatest Elong: 45.4°E
06	23:07	Moon at Perigee: 364860 km
09	20:19	Saturn 3.0°N of Moon
10	19:31	LAST QUARTER MOON
14	06:33	Jupiter 1.5°S of Moon
16	00:47	Pleiades 1.8°N of Moon
18	04:37	NEW MOON
20	09:10	Pollux 1.7°N of Moon
21	14:58	Summer Solstice
22	00:47	Venus 3.7°S of Moon
22	10:09	Mars 3.8°S of Moon
22	18:30	Moon at Apogee: 405385 km
26	07:50	FIRST QUARTER MOON

11

Date	GMT (h:m)	Event
Jul 01	05	Mercury at Superior Conjunction
01	06	Venus 3.6°N of Mars
01	07:20	Antares 1.5°S of Moon
03	11:39	FULL MOON
04	22:28	Moon at Perigee: 360151 km
06	19	Earth at Aphelion: 1.01668 AU
07	03:05	Saturn 2.7°N of Moon
10	01:48	LAST QUARTER MOON
10	05:21	Mars 0.6°N of Regulus
11	21:18	Jupiter 2.2°S of Moon
13	06:31	Pleiades 1.7°N of Moon
16	08:05	Venus 1.7°S of Regulus
17	18:32	NEW MOON
19	08:56	Mercury 3.5°S of Moon
20	06:56	Moon at Apogee: 406291 km
21	04:00	Mars 3.3°S of Moon
25	03:01	Spica 2.8°S of Moon
25	22:07	FIRST QUARTER MOON
28	16	Delta-Aquarid Meteor Shower
28	17:11	Antares 1.3°S of Moon
28	18:21	Mercury 0.1°S of Regulus
Aug 01	18:31	FULL MOON
02	05:52	Moon at Perigee: 357311 km
03	10:21	Saturn 2.5°N of Moon
08	09:41	Jupiter 2.9°S of Moon
08	10:28	LAST QUARTER MOON
09	12:16	Pleiades 1.5°N of Moon
10	02	Mercury at Greatest Elong: 27.4°E
13	07	Perseid Meteor Shower
13	08	Mercury 4.7°S of Mars
13	11	Venus at Inferior Conjunction
13	21:36	Pollux 1.7°N of Moon
16	09:38	NEW MOON
16	11:55	Moon at Apogee: 406635 km
18	23:06	Mars 2.2°S of Moon
21	09:28	Spica 2.6°S of Moon
24	09:57	FIRST QUARTER MOON
25	01:29	Antares 1.1°S of Moon
27	08	Saturn at Opposition
30	15:51	Moon at Perigee: 357182 km
30	18:03	Saturn 2.5°N of Moon
31	01:35	FULL MOON
Sep 04	19:44	Jupiter 3.3°S of Moon
05	19:25	Pleiades 1.2°N of Moon
06	11	Mercury at Inferior Conjunction
06	22:21	LAST QUARTER MOON
10	03:32	Pollux 1.5°N of Moon
12	15:42	Moon at Apogee: 406289 km
15	01:40	NEW MOON
16	19:19	Mars 0.7°S of Moon: Occn.
17	15:12	Spica 2.4°S of Moon
21	07:50	Antares 0.9°S of Moon
22	13	Mercury at Greatest Elong: 17.9°W
22	19:32	FIRST QUARTER MOON
23	06:50	Autumnal Equinox
27	01:25	Saturn 2.7°N of Moon
28	01:05	Moon at Perigee: 359911 km
29	09:57	FULL MOON

Date	GMT (h:m)	Event
Oct 02	03:16	Jupiter 3.4°S of Moon
03	04:25	Pleiades 1.1°N of Moon
06	13:48	LAST QUARTER MOON
07	10:23	Pollux 1.4°N of Moon
09	06:08	Venus 2.3°S of Regulus
10	03:41	Moon at Apogee: 405426 km
14	17:55	NEW MOON
14	17:59	Annular Solar Eclipse; mag=0.952
18	13:17	Antares 0.9°S of Moon
20	05	Mercury at Superior Conjunction
22	00	Orionid Meteor Shower
22	03:29	FIRST QUARTER MOON
23	22	Venus at Greatest Elong: 46.4°W
24	07:52	Saturn 2.8°N of Moon
26	02:53	Moon at Perigee: 364873 km
28	20:14	Partial Lunar Eclipse; mag=0.122
28	20:24	FULL MOON
29	08:10	Jupiter 3.1°S of Moon
30	14:30	Pleiades 1.1°N of Moon
Nov 03	04	Jupiter at Opposition
03	18:31	Pollux 1.4°N of Moon
05	08:37	LAST QUARTER MOON
06	21:49	Moon at Apogee: 404569 km
09	09:28	Venus 1.0°S of Moon: Occn.
11	05:09	Spica 2.4°S of Moon
13	00	N Taurid Meteor Shower
13	09:27	NEW MOON
13	18	Uranus at Opposition
14	19:42	Antares 0.9°S of Moon
16	21:17	Mercury 2.5°N of Antares
18	05	Mars in Conjunction with Sun
18	06	Leonid Meteor Shower
20	10:50	FIRST QUARTER MOON
20	14:02	Saturn 2.7°N of Moon
21	21:03	Moon at Perigee: 369824 km
25	11:10	Jupiter 2.8°S of Moon
27	00:02	Pleiades 1.1°N of Moon
27	09:16	FULL MOON
Dec 01	03:23	Pollux 1.6°N of Moon
04	00:38	Regulus 4.0°S of Moon
04	14	Mercury at Greatest Elong: 21.3°E
04	18:42	Moon at Apogee: 404348 km
05	05:49	LAST QUARTER MOON
08	14:05	Spica 2.3°S of Moon
09	16:53	Venus 3.6°N of Moon
12	23:32	NEW MOON
14	19	Geminid Meteor Shower
16	18:53	Moon at Perigee: 367900 km
17	21:58	Saturn 2.5°N of Moon
19	18:39	FIRST QUARTER MOON
22	03:28	Winter Solstice
22	14:20	Jupiter 2.6°S of Moon
22	19	Mercury at Inferior Conjunction
23	03	Ursid Meteor Shower
24	07:37	Pleiades 1.1°N of Moon
27	00:33	FULL MOON
28	11:51	Pollux 1.7°N of Moon
31	08:52	Regulus 3.8°S of Moon

Almanac of Astronomical Events for 2024

Date	GMT (h:m)	Event
Jan 01	15:28	Moon at Apogee: 404911 km
03	01	Earth at Perihelion: 0.98330 AU
04	03:30	LAST QUARTER MOON
04	09	Quadrantid Meteor Shower
04	23:06	Spica 2.0°S of Moon
08	14:24	Antares 0.8°S of Moon
10	08:31	Mars 4.2°N of Moon
11	11:57	NEW MOON
12	14	Mercury at Greatest Elong: 23.5°W
13	10:35	Moon at Perigee: 362264 km
14	09:31	Saturn 2.1°N of Moon
18	03:53	FIRST QUARTER MOON
18	20:40	Jupiter 2.8°S of Moon
20	13:25	Pleiades 0.9°N of Moon
24	19:00	Pollux 1.7°N of Moon
25	17:54	FULL MOON
27	16	Mercury 0.2°N of Mars
27	16:18	Regulus 3.6°S of Moon
29	08:14	Moon at Apogee: 405781 km
Feb 01	07:04	Spica 1.7°S of Moon
02	23:18	LAST QUARTER MOON
05	00:15	Antares 0.6°S of Moon
08	06:30	Mars 4.2°N of Moon
09	22:59	NEW MOON
10	18:49	Moon at Perigee: 358088 km
11	00:37	Saturn 1.8°N of Moon
15	08:15	Jupiter 3.2°S of Moon
16	15:01	FIRST QUARTER MOON
16	19:13	Pleiades 0.6°N of Moon
21	00:54	Pollux 1.6°N of Moon
22	09	Venus 0.6°N of Mars
23	22:45	Regulus 3.6°S of Moon
24	12:30	FULL MOON
25	15:00	Moon at Apogee: 406316 km
28	08	Mercury at Superior Conjunction
28	13:40	Spica 1.5°S of Moon
28	21	Saturn in Conjunction with Sun
Mar 03	08:16	Antares 0.4°S of Moon
03	15:24	LAST QUARTER MOON
08	04:59	Mars 3.5°N of Moon
08	17:01	Venus 3.3°N of Moon
10	07:06	Moon at Perigee: 356895 km
10	09:00	NEW MOON
14	01:02	Jupiter 3.6°S of Moon
15	02:54	Pleiades 0.4°N of Moon
17	04:11	FIRST QUARTER MOON
19	06:44	Pollux 1.5°N of Moon
20	03:07	Vernal Equinox
21	22	Venus 0.3°N of Saturn
22	04:46	Regulus 3.6°S of Moon
23	15:44	Moon at Apogee: 406292 km
24	22	Mercury at Greatest Elong: 18.7°E
25	07:00	FULL MOON
25	07:13	Pen. Lunar Eclipse; mag=0.956
26	19:40	Spica 1.4°S of Moon
30	14:24	Antares 0.3°S of Moon

Date	GMT (h:m)	Event
Apr 02	03:15	LAST QUARTER MOON
06	03:51	Mars 2.0°N of Moon
06	09:20	Saturn 1.2°N of Moon: Occn.
07	16:39	Venus 0.4°S of Moon: Occn.
07	17:53	Moon at Perigee: 358850 km
08	18:17	Total Solar Eclipse; mag=1.057
08	18:21	NEW MOON
10	19	Mars 0.4°N of Saturn
10	21:08	Jupiter 4.0°S of Moon
11	12:38	Pleiades 0.4°N of Moon
11	23	Mercury at Inferior Conjunction
15	13:47	Pollux 1.5°N of Moon
15	19:13	FIRST QUARTER MOON
18	11:14	Regulus 3.6°S of Moon
20	02:09	Moon at Apogee: 405625 km
22	07	Lyrid Meteor Shower
23	02:02	Spica 1.5°S of Moon
23	23:49	FULL MOON
26	20:00	Antares 0.3°S of Moon
May 01	11:27	LAST QUARTER MOON
03	22:26	Saturn 0.8°N of Moon: Occn.
04	20	Eta-Aquarid Meteor Shower
05	02:26	Mars 0.2°S of Moon: Occn.
05	22:11	Moon at Perigee: 363166 km
06	08:25	Mercury 3.8°S of Moon
08	03:22	NEW MOON
08	11	Mars at Perihelion: 1.38150 AU
09	21	Mercury at Greatest Elong: 26.4°W
12	22:17	Pollux 1.6°N of Moon
13	11	Uranus in Conjunction with Sun
15	11:48	FIRST QUARTER MOON
15	18:43	Regulus 3.5°S of Moon
17	19:00	Moon at Apogee: 404641 km
18	18	Jupiter in Conjunction with Sun
20	09:20	Spica 1.4°S of Moon
23	13:53	FULL MOON
24	02:31	Antares 0.4°S of Moon
30	17:13	LAST QUARTER MOON
31	08:01	Saturn 0.4°N of Moon: Occn.
Jun 02	07:23	Moon at Perigee: 368108 km
02	23:37	Mars 2.4°S of Moon
04	15	Venus at Superior Conjunction
05	08:14	Pleiades 0.4°N of Moon
06	12:38	NEW MOON
09	07:23	Pollux 1.7°N of Moon
12	03:00	Regulus 3.3°S of Moon
14	05:18	FIRST QUARTER MOON
14	13:36	Moon at Apogee: 404078 km
14	16	Mercury at Superior Conjunction
16	17:28	Spica 1.2°S of Moon
20	10:33	Antares 0.3°S of Moon
20	20:51	Summer Solstice
22	01:08	FULL MOON
27	11:45	Moon at Perigee: 369292 km
27	14:52	Saturn 0.1°S of Moon: Occn.
28	21:53	LAST QUARTER MOON

Date	GMT (h:m)	Event
Jul 01	18:27	Mars 4.1°S of Moon
02	15:30	Pleiades 0.3°N of Moon
05	05	Earth at Aphelion: 1.01673 AU
05	22:57	NEW MOON
07	18:33	Mercury 3.2°S of Moon
09	11:20	Regulus 3.1°S of Moon
12	08:12	Moon at Apogee: 404363 km
13	22:49	FIRST QUARTER MOON
14	01:48	Spica 0.9°S of Moon
17	19:37	Antares 0.2°S of Moon
21	10:17	FULL MOON
22	07	Mercury at Greatest Elong: 26.9°E
24	05:43	Moon at Perigee: 364914 km
24	20:38	Saturn 0.4°S of Moon: Occn.
25	01:38	Mercury 1.7°S of Regulus
27	22	Delta-Aquarid Meteor Shower
28	02:51	LAST QUARTER MOON
29	21:13	Pleiades 0.1°N of Moon
Aug 02	22:58	Pollux 1.8°N of Moon
04	11:13	NEW MOON
04	22:51	Venus 1.0°N of Regulus
05	18:54	Regulus 2.9°S of Moon
05	22:04	Venus 1.7°S of Moon
09	01:32	Moon at Apogee: 405298 km
10	09:34	Spica 0.7°S of Moon
12	14	Perseid Meteor Shower
12	15:19	FIRST QUARTER MOON
14	04:38	Antares 0.0°S of Moon
14	15	Mars 0.3°N of Jupiter
19	02	Mercury at Inferior Conjunction
19	18:26	FULL MOON
21	02:54	Saturn 0.4°S of Moon: Occn.
21	05:05	Moon at Perigee: 360199 km
26	02:54	Pleiades 0.1°S of Moon
26	09:26	LAST QUARTER MOON
30	04:47	Pollux 1.7°N of Moon
Sep 03	01:55	NEW MOON
05	02	Mercury at Greatest Elong: 18.1°W
05	10:13	Venus 1.2°N of Moon
05	14:55	Moon at Apogee: 406215 km
06	16:22	Spica 0.6°S of Moon
08	04	Saturn at Opposition
09	02:50	Mercury 0.4°N of Regulus
10	12:29	Antares 0.1°N of Moon
11	06:06	FIRST QUARTER MOON
17	10:14	Saturn 0.3°S of Moon: Occn.
18	02:34	FULL MOON
18	02:44	Partial Lunar Eclipse; mag=0.085
18	02:57	Venus 2.2°N of Spica
18	13:26	Moon at Perigee: 357284 km
20	23	Neptune at Opposition
22	10:17	Pleiades 0.2°S of Moon
22	12:44	Autumnal Equinox
24	18:50	LAST QUARTER MOON
26	10:25	Pollux 1.6°N of Moon
29	07:16	Regulus 3.0°S of Moon
30	21	Mercury at Superior Conjunction

Date	GMT (h:m)	Event
Oct 02	18:45	Annular Solar Eclipse; mag=0.933
02	18:49	NEW MOON
02	19:40	Moon at Apogee: 406517 km
05	20:27	Venus 3.0°N of Moon
07	18:48	Antares 0.2°N of Moon
10	18:55	FIRST QUARTER MOON
14	18:05	Saturn 0.1°S of Moon: Occn.
17	00:46	Moon at Perigee: 357173 km
17	11:26	FULL MOON
19	19:59	Pleiades 0.1°S of Moon
21	06	Orionid Meteor Shower
23	17:16	Pollux 1.7°N of Moon
23	19:55	Mars 3.9°S of Moon
24	08:03	LAST QUARTER MOON
25	23:43	Venus 3.0°N of Antares
26	13:24	Regulus 2.9°S of Moon
29	22:50	Moon at Apogee: 406164 km
Nov 01	12:47	NEW MOON
03	07:37	Mercury 2.1°N of Moon
04	00:26	Antares 0.1°N of Moon
05	00:16	Venus 3.1°N of Moon
05	06	S Taurid Meteor Shower
09	05:56	FIRST QUARTER MOON
10	04:22	Mercury 2.0°N of Antares
11	01:36	Saturn 0.1°S of Moon: Occn.
12	06	N Taurid Meteor Shower
14	11:18	Moon at Perigee: 360110 km
15	21:29	FULL MOON
16	06:59	Pleiades 0.1°S of Moon
16	08	Mercury at Greatest Elong: 22.5°E
17	03	Uranus at Opposition
17	12	Leonid Meteor Shower
20	02:07	Pollux 1.9°N of Moon
20	21:07	Mars 2.4°S of Moon
22	20:48	Regulus 2.7°S of Moon
23	01:28	LAST QUARTER MOON
26	11:56	Moon at Apogee: 405315 km
27	11:33	Spica 0.4°S of Moon
Dec 01	06:21	NEW MOON
04	22:40	Venus 2.3°N of Moon
06	02	Mercury at Inferior Conjunction
07	20	Jupiter at Opposition
08	08:49	Saturn 0.3°S of Moon: Occn.
08	15:27	FIRST QUARTER MOON
12	13:18	Moon at Perigee: 365360 km
13	17:13	Pleiades 0.1°S of Moon
14	01	Geminid Meteor Shower
15	09:02	FULL MOON
17	12:12	Pollux 2.0°N of Moon
18	08:46	Mars 0.9°S of Moon: Occn.
20	05:37	Regulus 2.5°S of Moon
21	09:20	Winter Solstice
22	10	Ursid Meteor Shower
22	22:18	LAST QUARTER MOON
24	07:25	Moon at Apogee: 404486 km
24	19:28	Spica 0.2°S of Moon
25	02	Mercury at Greatest Elong: 22.0°W
28	14:37	Antares 0.1°N of Moon
30	22:27	NEW MOON

Almanac of Astronomical Events for 2025

Date	GMT (h:m)	Event
Jan 03	15:24	Venus 1.4°N of Moon
03	15	Quadrantid Meteor Shower
04	14	Earth at Perihelion: 0.98333 AU
04	17:18	Saturn 0.7°S of Moon: Occn.
06	23:56	FIRST QUARTER MOON
07	23:34	Moon at Perigee: 370173 km
10	01:01	Pleiades 0.3°S of Moon
10	04	Venus at Greatest Elong: 47.2°E
13	21:45	Pollux 2.1°N of Moon
13	22:27	FULL MOON
14	03:42	Mars 0.2°S of Moon: Occn.
16	01	Mars at Opposition
16	14:57	Regulus 2.2°S of Moon
18	16	Venus 2.2°N of Saturn
21	03:53	Spica 0.1°N of Moon
21	04:55	Moon at Apogee: 404299 km
21	20:31	LAST QUARTER MOON
23	17:07	Mars 2.3°S of Pollux
24	23:34	Antares 0.3°N of Moon
29	12:36	NEW MOON
Feb 01	04:46	Saturn 1.1°S of Moon: Occn.
01	20:27	Venus 2.3°N of Moon
02	02:43	Moon at Perigee: 367457 km
05	08:02	FIRST QUARTER MOON
06	06:43	Pleiades 0.5°S of Moon
09	12	Mercury at Superior Conjunction
09	19:36	Mars 0.8°S of Moon: Occn.
10	05:19	Pollux 2.1°N of Moon
12	13:53	FULL MOON
12	23:21	Regulus 2.2°S of Moon
17	12:01	Spica 0.3°N of Moon
18	01:11	Moon at Apogee: 404882 km
20	17:33	LAST QUARTER MOON
21	08:21	Antares 0.4°N of Moon
28	00:45	NEW MOON
Mar 01	04:03	Mercury 0.4°N of Moon: Occn.
01	21:18	Moon at Perigee: 361967 km
05	12:32	Pleiades 0.6°S of Moon
06	16:32	FIRST QUARTER MOON
08	06	Mercury at Greatest Elong: 18.2°E
09	00:27	Mars 1.7°S of Moon
09	11:06	Pollux 2.0°N of Moon
12	06:07	Regulus 2.2°S of Moon
12	10	Saturn in Conjunction with Sun
14	06:55	FULL MOON
14	06:59	Total Lunar Eclipse; mag=1.178
16	19:16	Spica 0.3°N of Moon
17	16:37	Moon at Apogee: 405754 km
20	09:02	Vernal Equinox
20	15:58	Antares 0.5°N of Moon
22	11:30	LAST QUARTER MOON
23	01	Venus at Inferior Conjunction
24	20	Mercury at Inferior Conjunction
29	10:47	Partial Solar Eclipse; mag=0.938
29	10:58	NEW MOON
29	19:29	Mars 3.9°S of Pollux

Date	GMT (h:m)	Event
Mar 29	19:29	Mars 3.9°S of Pollux
30	05:26	Moon at Perigee: 358127 km
Apr 01	20:28	Pleiades 0.6°S of Moon
05	02:15	FIRST QUARTER MOON
05	16:46	Pollux 2.0°N of Moon
05	19:04	Mars 2.2°S of Moon
08	11:51	Regulus 2.2°S of Moon
10	12	Mercury 2.1°N of Saturn
13	00:22	FULL MOON
13	01:39	Spica 0.3°N of Moon
13	22:48	Moon at Apogee: 406295 km
16	22	Mars at Aphelion: 1.66606 AU
16	22:19	Antares 0.4°N of Moon
21	01:36	LAST QUARTER MOON
21	19	Mercury at Greatest Elong: 27.4°W
22	13	Lyrid Meteor Shower
25	01:21	Venus 2.4°N of Moon
25	04:15	Saturn 2.3°S of Moon
26	01:05	Mercury 4.4°S of Moon
27	16:15	Moon at Perigee: 357119 km
27	19:31	NEW MOON
28	19	Venus 3.7°N of Saturn
29	06:35	Pleiades 0.5°S of Moon
May 03	00:02	Pollux 2.1°N of Moon
03	23:12	Mars 2.1°S of Moon
04	13:52	FIRST QUARTER MOON
05	02	Eta-Aquarid Meteor Shower
05	17:58	Regulus 2.0°S of Moon
10	07:43	Spica 0.4°N of Moon
11	00:49	Moon at Apogee: 406245 km
12	16:56	FULL MOON
14	04:10	Antares 0.3°N of Moon
18	01	Uranus in Conjunction with Sun
20	11:59	LAST QUARTER MOON
22	17:51	Saturn 2.8°S of Moon
23	23:52	Venus 4.0°S of Moon
26	01:37	Moon at Perigee: 359023 km
27	03:02	NEW MOON
30	04	Mercury at Superior Conjunction
30	09:13	Pollux 2.3°N of Moon
Jun 01	02	Venus at Greatest Elong: 45.9°W
01	09:49	Mars 1.4°S of Moon
02	01:30	Regulus 1.8°S of Moon
03	03:41	FIRST QUARTER MOON
06	14:15	Spica 0.5°N of Moon
07	10:42	Moon at Apogee: 405553 km
10	10:25	Antares 0.3°N of Moon
11	07:44	FULL MOON
17	02:05	Mars 0.7°N of Regulus
18	19:19	LAST QUARTER MOON
19	03:47	Saturn 3.4°S of Moon
21	02:42	Summer Solstice
23	02:59	Pleiades 0.6°S of Moon
23	04:43	Moon at Perigee: 363178 km
24	15	Jupiter in Conjunction with Sun

Jun 25	10:31	NEW MOON
26	19:14	Pollux 2.5°N of Moon
27	06:02	Mercury 2.9°S of Moon
29	10:26	Regulus 1.5°S of Moon
30	01:05	Mars 0.2°S of Moon: Occn.
Jul 02	19:30	FIRST QUARTER MOON
03	21	Earth at Aphelion: 1.01664 AU
03	21:39	Spica 0.8°N of Moon
04	04	Mercury at Greatest Elong: 25.9°E
05	02:29	Moon at Apogee: 404627 km
07	17:37	Antares 0.4°N of Moon
10	20:37	FULL MOON
13	08:32	Venus 3.1°N of Aldebaran
16	10:19	Saturn 3.8°S of Moon
18	00:38	LAST QUARTER MOON
20	10:27	Pleiades 0.7°S of Moon
20	13:52	Moon at Perigee: 368047 km
24	19:11	NEW MOON
26	19:44	Regulus 1.4°S of Moon
28	04	Delta-Aquarid Meteor Shower
28	19:45	Mars 1.3°N of Moon
31	05:45	Spica 1.0°N of Moon
Aug 01	00	Mercury at Inferior Conjunction
01	12:41	FIRST QUARTER MOON
01	20:37	Moon at Apogee: 404164 km
04	01:40	Antares 0.6°N of Moon
09	07:55	FULL MOON
12	07	Venus 0.9°S of Jupiter
12	15:05	Saturn 4.0°S of Moon
12	20	Perseid Meteor Shower
14	18:01	Moon at Perigee: 369287 km
16	05:12	LAST QUARTER MOON
16	16:09	Pleiades 0.9°S of Moon
19	10	Mercury at Greatest Elong: 18.6°W
19	21:05	Jupiter 4.8°S of Moon
20	12:07	Pollux 2.4°N of Moon
21	16:14	Mercury 3.7°S of Moon
23	06:06	NEW MOON
26	16:41	Mars 2.8°N of Moon
27	13:57	Spica 1.1°N of Moon
29	15:34	Moon at Apogee: 404552 km
31	06:25	FIRST QUARTER MOON
31	09:55	Antares 0.7°N of Moon
Sep 07	18:09	FULL MOON
07	18:12	Total Lunar Eclipse; mag=1.362
08	20:09	Saturn 4.0°S of Moon
10	12:09	Moon at Perigee: 364781 km
12	21:48	Pleiades 1.0°S of Moon
13	03:28	Mars 2.0°N of Spica
13	11	Mercury at Superior Conjunction
14	10:33	LAST QUARTER MOON
16	11:06	Jupiter 4.6°S of Moon
16	17:58	Pollux 2.4°N of Moon
19	08:57	Venus 0.4°N of Regulus
19	11:11	Regulus 1.3°S of Moon
19	11:46	Venus 0.8°S of Moon: Occn.
21	05	Saturn at Opposition
21	19:42	Partial Solar Eclipse; mag=0.855
21	19:54	NEW MOON
22	18:20	Autumnal Equinox
23	11	Neptune at Opposition

Sep 23	21:31	Spica 1.1°N of Moon
24	14:50	Mars 3.9°N of Moon
26	09:46	Moon at Apogee: 405552 km
27	17:34	Antares 0.6°N of Moon
29	23:54	FIRST QUARTER MOON
Oct 06	02:46	Saturn 3.8°S of Moon
07	03:47	FULL MOON
08	12:36	Moon at Perigee: 359819 km
10	05:20	Pleiades 0.9°S of Moon
13	18:13	LAST QUARTER MOON
13	22:31	Jupiter 4.3°S of Moon
13	23:31	Pollux 2.5°N of Moon
16	16:56	Regulus 1.3°S of Moon
19	20	Mercury 2.0°S of Mars
19	21:37	Venus 3.7°N of Moon
21	12	Orionid Meteor Shower
21	12:25	NEW MOON
23	16:15	Mercury 2.3°N of Moon
23	23:31	Moon at Apogee: 406445 km
25	00:15	Antares 0.5°N of Moon
29	16:21	FIRST QUARTER MOON
29	22	Mercury at Greatest Elong: 23.9°E
Nov 02	01:02	Venus 3.3°N of Spica
02	10:46	Saturn 3.7°S of Moon
05	13:19	FULL MOON
05	22:29	Moon at Perigee: 356833 km
06	15:26	Pleiades 0.8°S of Moon
09	02:41	Mercury 2.6°N of Antares
10	06:40	Pollux 2.7°N of Moon
10	07:56	Jupiter 4.0°S of Moon
12	05:28	LAST QUARTER MOON
12	22:51	Regulus 1.1°S of Moon
13	04	Mercury 1.2°S of Mars
17	10:11	Spica 1.2°N of Moon
17	18	Leonid Meteor Shower
20	02:48	Moon at Apogee: 406693 km
20	06:47	NEW MOON
20	09	Mercury at Inferior Conjunction
21	13	Uranus at Opposition
28	06:59	FIRST QUARTER MOON
29	19:08	Saturn 3.7°S of Moon
Dec 04	02:54	Pleiades 0.8°S of Moon
04	11:06	Moon at Perigee: 356962 km
04	23:14	FULL MOON
07	15:48	Jupiter 3.7°S of Moon
07	16:21	Pollux 2.9°N of Moon
07	21	Mercury at Greatest Elong: 20.7°W
10	06:32	Regulus 0.8°S of Moon
11	20:52	LAST QUARTER MOON
14	07	Geminid Meteor Shower
14	16:27	Spica 1.4°N of Moon
17	06:09	Moon at Apogee: 406324 km
18	12:29	Antares 0.4°N of Moon
20	01:43	NEW MOON
21	15:03	Winter Solstice
22	16	Ursid Meteor Shower
27	03:24	Saturn 4.0°S of Moon
27	19:10	FIRST QUARTER MOON
31	13:21	Pleiades 0.9°S of Moon

16

Almanac of Astronomical Events for 2026

Date	GMT (h:m)	Event
Jan 01	21:43	Moon at Perigee: 360348 km
03	10:03	FULL MOON
03	17	Earth at Perihelion: 0.98330 AU
03	22	Quadrantid Meteor Shower
03	22:01	Jupiter 3.7°S of Moon
04	03:28	Pollux 3.0°N of Moon
06	16	Venus at Superior Conjunction
06	16:20	Regulus 0.5°S of Moon
09	10	Mars in Conjunction with Sun
10	08	Jupiter at Opposition
10	15:48	LAST QUARTER MOON
10	23:50	Spica 1.6°N of Moon
13	20:48	Moon at Apogee: 405437 km
14	19:28	Antares 0.6°N of Moon
18	19:52	NEW MOON
21	16	Mercury at Superior Conjunction
23	12:31	Saturn 4.3°S of Moon
26	04:47	FIRST QUARTER MOON
27	21:07	Pleiades 1.1°S of Moon
29	21:53	Moon at Perigee: 365878 km
31	02:31	Jupiter 3.8°S of Moon
31	13:45	Pollux 3.0°N of Moon
Feb 01	22:09	FULL MOON
03	02:48	Regulus 0.4°S of Moon
07	08:26	Spica 1.8°N of Moon
09	12:43	LAST QUARTER MOON
10	16:52	Moon at Apogee: 404577 km
11	03:19	Antares 0.7°N of Moon
17	12:01	NEW MOON
17	12:12	Annular Solar Eclipse; mag=0.963
18	23:03	Mercury 0.1°N of Moon: Occn.
19	18	Mercury at Greatest Elong: 18.1°E
19	23:54	Saturn 4.6°S of Moon
24	02:43	Pleiades 1.2°S of Moon
24	12:28	FIRST QUARTER MOON
24	23:18	Moon at Perigee: 370132 km
27	06:26	Jupiter 4.0°S of Moon
27	21:34	Pollux 3.0°N of Moon
Mar 02	12:00	Regulus 0.4°S of Moon
03	11:34	Total Lunar Eclipse; mag=1.151
03	11:38	FULL MOON
06	17:24	Spica 1.8°N of Moon
07	11	Mercury at Inferior Conjunction
10	11:32	Antares 0.7°N of Moon
10	13:43	Moon at Apogee: 404385 km
11	09:39	LAST QUARTER MOON
15	19	Mercury 3.4°N of Mars
17	14:07	Mercury 2.0°N of Moon
17	21:51	Mars 1.5°S of Moon
19	01:23	NEW MOON
20	12:39	Venus 4.6°S of Moon
20	14:46	Vernal Equinox
22	10	Neptune in Conjunction with Sun
22	11:40	Moon at Perigee: 366858 km
23	08:32	Pleiades 1.1°S of Moon
25	08	Saturn in Conjunction with Sun
25	19:18	FIRST QUARTER MOON

Date	GMT (h:m)	Event
Mar 26	07	Mars at Perihelion: 1.38126 AU
26	12:13	Jupiter 3.9°S of Moon
27	03:18	Pollux 3.0°N of Moon
29	19:00	Regulus 0.4°S of Moon
Apr 02	02:12	FULL MOON
03	01:32	Spica 1.8°N of Moon
03	23	Mercury at Greatest Elong: 27.8°W
06	19:21	Antares 0.6°N of Moon
07	08:32	Moon at Apogee: 404974 km
10	04:52	LAST QUARTER MOON
16	00:45	Mars 3.7°S of Moon
17	11:52	NEW MOON
19	06:57	Moon at Perigee: 361631 km
19	08:49	Venus 4.8°S of Moon
19	16:28	Pleiades 1.0°S of Moon
19	19	Mars 1.2°N of Saturn
20	10	Mercury 0.5°S of Saturn
20	22	Mercury 1.7°S of Mars
22	19	Lyrid Meteor Shower
22	22:06	Jupiter 3.6°S of Moon
23	08:59	Pollux 3.2°N of Moon
24	02:32	FIRST QUARTER MOON
24	04:17	Venus 3.4°S of Pleiades
26	00:37	Regulus 0.2°S of Moon
30	08:17	Spica 1.8°N of Moon
May 01	17:23	FULL MOON
04	02:20	Antares 0.5°N of Moon
04	22:30	Moon at Apogee: 405843 km
05	08	Eta-Aquarid Meteor Shower
09	21:10	LAST QUARTER MOON
14	14	Mercury at Superior Conjunction
16	20:01	NEW MOON
17	13:48	Moon at Perigee: 358074 km
19	01:50	Venus 2.9°S of Moon
20	12:39	Jupiter 3.1°S of Moon
20	16:30	Pollux 3.4°N of Moon
22	16	Uranus in Conjunction with Sun
23	06:41	Regulus 0.0°N of Moon
23	11:11	FIRST QUARTER MOON
27	14:09	Spica 1.9°N of Moon
31	08:32	Antares 0.4°N of Moon
31	08:45	FULL MOON
Jun 01	04:32	Moon at Apogee: 406369 km
08	10:00	LAST QUARTER MOON
09	20	Venus 1.6°N of Jupiter
13	13:15	Pleiades 1.0°S of Moon
14	23:18	Moon at Perigee: 357196 km
15	02:54	NEW MOON
15	20	Mercury at Greatest Elong: 24.5°E
16	19:32	Mercury 2.6°S of Moon
17	02:08	Pollux 3.6°N of Moon
17	06:54	Jupiter 2.5°S of Moon
17	20:21	Venus 0.3°S of Moon: Occn.
19	14:31	Regulus 0.3°N of Moon
21	08:25	Summer Solstice
21	21:55	FIRST QUARTER MOON

17

Date	GMT (h:m)	Event
Jun 23	20:11	Spica 2.2°N of Moon
25	12	Mercury 3.8°S of Jupiter
27	14:32	Antares 0.5°N of Moon
28	07:11	Moon at Apogee: 406267 km
29	23:57	FULL MOON
Jul 06	18	Earth at Aphelion: 1.01664 AU
07	19:29	LAST QUARTER MOON
09	14:36	Venus 0.9°N of Regulus
10	22:54	Pleiades 1.1°S of Moon
13	01	Mercury at Inferior Conjunction
13	07:50	Moon at Perigee: 359111 km
14	09:43	NEW MOON
17	00:07	Regulus 0.5°N of Moon
17	16:31	Venus 2.0°N of Moon
21	03:21	Spica 2.4°N of Moon
21	11:06	FIRST QUARTER MOON
24	21:00	Antares 0.6°N of Moon
25	16:45	Moon at Apogee: 405549 km
28	10	Delta-Aquarid Meteor Shower
29	12	Jupiter in Conjunction with Sun
29	14:36	FULL MOON
Aug 02	08	Mercury at Greatest Elong: 19.5°W
06	02:21	LAST QUARTER MOON
07	06:23	Pleiades 1.2°S of Moon
09	05:31	Mars 4.4°S of Moon
10	11:18	Moon at Perigee: 363288 km
10	22:38	Pollux 3.6°N of Moon
11	12:48	Mercury 2.1°S of Moon
12	17:37	NEW MOON
12	17:46	Total Solar Eclipse; mag=1.039
13	02	Perseid Meteor Shower
15	06	Venus at Greatest Elong: 45.9°E
16	08:47	Venus 2.1°N of Moon
17	11:49	Spica 2.4°N of Moon
20	02:46	FIRST QUARTER MOON
21	04:18	Antares 0.6°N of Moon
22	08:20	Moon at Apogee: 404644 km
27	17	Mercury at Superior Conjunction
28	04:13	Partial Lunar Eclipse; mag=0.930
28	04:18	FULL MOON
Sep 01	13:24	Venus 1.2°S of Spica
03	12:03	Pleiades 1.2°S of Moon
04	07:51	LAST QUARTER MOON
06	18:24	Mars 3.0°S of Moon
06	20:26	Moon at Perigee: 368255 km
07	06:32	Pollux 3.6°N of Moon
08	18:13	Jupiter 0.8°S of Moon: Occn.
09	19:36	Regulus 0.5°N of Moon
11	03:27	NEW MOON
13	20:53	Spica 2.4°N of Moon
14	11:10	Venus 0.5°S of Moon: Occn.
17	12:18	Antares 0.6°N of Moon
18	20:44	FIRST QUARTER MOON
19	03:00	Moon at Apogee: 404217 km
23	00:06	Autumnal Equinox
26	00	Neptune at Opposition
26	01:49	Mercury 0.8°N of Spica
26	16:49	FULL MOON
30	17:39	Pleiades 1.1°S of Moon

Date	GMT (h:m)	Event
Oct 01	20:41	Moon at Perigee: 369338 km
03	13:25	LAST QUARTER MOON
04	12	Saturn at Opposition
04	12:27	Pollux 3.8°N of Moon
05	05:30	Mars 1.2°S of Moon: Occn.
06	10:18	Jupiter 0.2°S of Moon: Occn.
07	02:57	Regulus 0.6°N of Moon
10	15:50	NEW MOON
12	02:30	Venus 3.1°S of Moon
12	10	Mercury at Greatest Elong: 25.2°E
12	20:08	Mercury 2.1°N of Moon
14	20:25	Antares 0.4°N of Moon
16	22:56	Moon at Apogee: 404639 km
18	16:13	FIRST QUARTER MOON
21	18	Orionid Meteor Shower
24	03	Venus at Inferior Conjunction
26	04:12	FULL MOON
28	01:11	Pleiades 1.0°S of Moon
28	18:01	Moon at Perigee: 364411 km
31	18:00	Pollux 4.0°N of Moon
Nov 01	20:28	LAST QUARTER MOON
02	14:23	Mars 1.1°N of Moon: Occn.
02	23:11	Jupiter 0.5°N of Moon: Occn.
03	08:40	Regulus 0.8°N of Moon
04	14	Mercury at Inferior Conjunction
07	11:31	Venus 1.1°N of Moon: Occn.
07	12:40	Spica 2.4°N of Moon
09	07:02	NEW MOON
10	13:49	Venus 0.1°S of Spica
11	03:58	Antares 0.3°N of Moon
13	17:50	Moon at Apogee: 405619 km
16	04	Mars 1.2°N of Jupiter
17	11:48	FIRST QUARTER MOON
18	00	Leonid Meteor Shower
20	23	Mercury at Greatest Elong: 19.6°W
24	11:18	Pleiades 0.9°S of Moon
24	14:53	FULL MOON
25	07:47	Mars 1.6°N of Regulus
25	20:58	Moon at Perigee: 359348 km
25	23	Uranus at Opposition
28	01:27	Pollux 4.2°N of Moon
30	09:18	Jupiter 1.2°N of Moon: Occn.
30	14:35	Regulus 1.1°N of Moon
30	19:32	Mars 3.3°N of Moon
Dec 01	06:09	LAST QUARTER MOON
04	18:36	Spica 2.5°N of Moon
09	00:52	NEW MOON
11	06:46	Moon at Apogee: 406421 km
12	15:35	Jupiter 1.3°N of Regulus
14	13	Geminid Meteor Shower
17	05:43	FIRST QUARTER MOON
21	20:50	Winter Solstice
21	22:37	Pleiades 1.0°S of Moon
22	22	Ursid Meteor Shower
24	01:28	FULL MOON
24	08:30	Moon at Perigee: 356650 km
27	17:32	Jupiter 1.5°N of Moon
27	22:44	Regulus 1.4°N of Moon
30	18:59	LAST QUARTER MOON

Almanac of Astronomical Events for 2027

Date	GMT (h:m)	Event
Jan 01	00:29	Spica 2.8°N of Moon
01	17	Mercury at Superior Conjunction
03	02	Earth at Perihelion: 0.98333 AU
03	19	Venus at Greatest Elong: 47.0°W
04	04	Quadrantid Meteor Shower
04	16:34	Antares 0.3°N of Moon
07	08:10	Moon at Apogee: 406610 km
07	20:24	NEW MOON
15	20:34	FIRST QUARTER MOON
18	08:46	Pleiades 1.1°S of Moon
21	21:49	Moon at Perigee: 357285 km
21	23:18	Pollux 4.4°N of Moon
22	12:17	FULL MOON
24	00:24	Jupiter 1.5°N of Moon
24	09:19	Regulus 1.5°N of Moon
28	07:55	Spica 2.9°N of Moon
29	10:55	LAST QUARTER MOON
31	22:48	Antares 0.4°N of Moon
Feb 03	06	Mercury at Greatest Elong: 18.3°E
03	13:31	Moon at Apogee: 406189 km
06	15:56	NEW MOON
06	16:00	Annular Solar Eclipse; mag=0.928
08	03:36	Mercury 0.4°N of Moon: Occn.
11	00	Jupiter at Opposition
14	07:58	FIRST QUARTER MOON
14	16:13	Pleiades 1.2°S of Moon
18	09:50	Pollux 4.4°N of Moon
18	17	Mercury at Inferior Conjunction
19	07:30	Moon at Perigee: 361015 km
19	15	Mars at Opposition
20	05:46	Jupiter 1.2°N of Moon: Occn.
20	20:33	Regulus 1.5°N of Moon
20	23:13	Pen. Lunar Eclipse; mag=0.927
20	23:23	FULL MOON
22	04:31	Mars 3.8°N of Regulus
24	17:14	Spica 2.9°N of Moon
28	05:16	LAST QUARTER MOON
28	06:09	Antares 0.4°N of Moon
Mar 03	05:41	Moon at Apogee: 405221 km
04	18:59	Venus 2.1°N of Moon
04	23	Mars at Aphelion: 1.66610 AU
06	05:28	Mercury 1.2°N of Moon
08	09:29	NEW MOON
13	21:43	Pleiades 1.1°S of Moon
15	16:25	FIRST QUARTER MOON
17	07	Mercury at Greatest Elong: 27.6°W
17	17:40	Pollux 4.5°N of Moon
19	04:31	Moon at Perigee: 366440 km
19	10:05	Jupiter 0.9°N of Moon: Occn.
19	18:33	Mars 4.3°N of Moon
20	06:13	Regulus 1.5°N of Moon
20	20:25	Vernal Equinox
22	10:44	FULL MOON
24	03:12	Spica 2.8°N of Moon
24	22	Neptune in Conjunction with Sun
27	14:36	Antares 0.3°N of Moon

Date	GMT (h:m)	Event
Mar 30	00:54	LAST QUARTER MOON
31	01:33	Moon at Apogee: 404331 km
31	06	Mars 4.3°N of Jupiter
Apr 04	06:49	Venus 3.3°S of Moon
06	23:51	NEW MOON
07	17	Saturn in Conjunction with Sun
10	03:30	Pleiades 0.9°S of Moon
13	22:57	FIRST QUARTER MOON
14	00:41	Moon at Perigee: 370001 km
15	15:01	Jupiter 1.0°N of Moon: Occn.
16	01:00	Mars 3.3°N of Moon
16	13:21	Regulus 1.6°N of Moon
20	12:09	Spica 2.8°N of Moon
20	22:27	FULL MOON
23	01	Lyrid Meteor Shower
23	23:16	Antares 0.1°N of Moon
27	21:21	Moon at Apogee: 404170 km
28	20:18	LAST QUARTER MOON
28	21	Mercury at Superior Conjunction
May 05	14	Eta-Aquarid Meteor Shower
06	10:58	NEW MOON
07	16	Venus 0.6°N of Saturn
09	20:07	Moon at Perigee: 366635 km
12	22:49	Jupiter 1.3°N of Moon
13	04:44	FIRST QUARTER MOON
13	18:57	Regulus 1.9°N of Moon
13	19:34	Mars 3.3°N of Moon
14	10:58	Mars 1.1°N of Regulus
17	19:16	Spica 2.9°N of Moon
20	10:59	FULL MOON
21	07:08	Antares 0.0°S of Moon
25	15:13	Moon at Apogee: 404801 km
27	08	Uranus in Conjunction with Sun
28	10	Mercury at Greatest Elong: 22.9°E
28	13:58	LAST QUARTER MOON
Jun 04	19:40	NEW MOON
06	06:29	Mercury 2.7°S of Moon
06	14:54	Moon at Perigee: 361703 km
09	10:54	Jupiter 1.9°N of Moon
10	01:01	Regulus 2.2°N of Moon
10	22:45	Mars 3.9°N of Moon
11	10:56	FIRST QUARTER MOON
14	01:01	Spica 3.0°N of Moon
17	13:46	Antares 0.0°S of Moon
19	00:44	FULL MOON
21	14:11	Summer Solstice
22	05:08	Moon at Apogee: 405698 km
23	08	Mercury at Inferior Conjunction
27	04:54	LAST QUARTER MOON

Date	GMT (h:m)	Event
Jul 01	07:04	Pleiades 0.9°S of Moon
04	03:02	NEW MOON
04	20:55	Moon at Perigee: 358284 km
05	06	Earth at Aphelion: 1.01673 AU
07	03:14	Jupiter 2.5°N of Moon
07	09:01	Regulus 2.3°N of Moon
09	07:52	Mars 4.7°N of Moon
10	18:39	FIRST QUARTER MOON
11	06:50	Spica 3.2°N of Moon
14	19:34	Antares 0.1°N of Moon
15	19	Mercury at Greatest Elong: 20.7°W
18	15:45	FULL MOON
18	16:03	Pen. Lunar Eclipse; mag=0.002
19	11:52	Moon at Apogee: 406217 km
25	06:23	Jupiter 0.4°N of Regulus
26	16:55	LAST QUARTER MOON
28	16	Delta-Aquarid Meteor Shower
28	16:20	Pleiades 1.0°S of Moon
Aug 02	06:28	Moon at Perigee: 357362 km
02	10:05	NEW MOON
02	10:07	Total Solar Eclipse; mag=1.079
03	19:00	Regulus 2.4°N of Moon
03	22:38	Jupiter 3.0°N of Moon
07	14:03	Spica 3.3°N of Moon
09	04:54	FIRST QUARTER MOON
11	01:31	Antares 0.1°N of Moon
11	11	Mercury at Superior Conjunction
12	00	Venus at Superior Conjunction
13	08	Perseid Meteor Shower
15	14:23	Moon at Apogee: 406086 km
17	07:14	Pen. Lunar Eclipse; mag=0.546
17	07:29	FULL MOON
23	14:36	Mars 1.7°N of Spica
24	23:36	Pleiades 0.9°S of Moon
25	02:27	LAST QUARTER MOON
30	15:36	Moon at Perigee: 359208 km
31	08	Jupiter in Conjunction with Sun
31	17:41	NEW MOON
Sep 02	04:46	Mercury 4.5°N of Moon
03	23:07	Spica 3.2°N of Moon
07	08:34	Antares 0.0°N of Moon
07	18:31	FIRST QUARTER MOON
11	23:39	Moon at Apogee: 405384 km
15	23:04	FULL MOON
21	03:21	Mercury 0.2°S of Spica
21	05:12	Pleiades 0.8°S of Moon
23	06:02	Autumnal Equinox
23	10:20	LAST QUARTER MOON
24	22	Mercury at Greatest Elong: 26.3°E
27	15:35	Regulus 2.4°N of Moon
27	20:12	Moon at Perigee: 363455 km
28	13	Neptune at Opposition
28	15:39	Jupiter 4.2°N of Moon
30	02:36	NEW MOON

Date	GMT (h:m)	Event
Oct 01	09:12	Spica 3.1°N of Moon
02	00:27	Mercury 1.8°N of Moon
03	08:58	Mars 4.5°N of Moon
04	17:01	Antares 0.2°S of Moon
07	11:47	FIRST QUARTER MOON
09	15:47	Moon at Apogee: 404531 km
11	14	Mercury 3.8°S of Venus
15	13:47	FULL MOON
18	00	Saturn at Opposition
18	10:48	Pleiades 0.6°S of Moon
19	15	Mercury at Inferior Conjunction
22	00	Orionid Meteor Shower
22	17:29	LAST QUARTER MOON
24	23:09	Regulus 2.6°N of Moon
25	05:36	Moon at Perigee: 368651 km
26	09:33	Jupiter 4.9°N of Moon
29	13:36	NEW MOON
29	13:47	Mars 3.7°N of Antares
31	07:57	Venus 4.4°N of Moon
Nov 01	02:13	Antares 0.4°S of Moon
01	06:53	Mars 3.3°N of Moon
04	09	Mercury at Greatest Elong: 18.8°W
06	01	S Taurid Meteor Shower
06	08:00	FIRST QUARTER MOON
06	11:37	Moon at Apogee: 404183 km
09	06:44	Venus 3.8°N of Antares
13	00	N Taurid Meteor Shower
14	03:26	FULL MOON
14	18:14	Pleiades 0.5°S of Moon
18	06	Leonid Meteor Shower
19	00:10	Moon at Perigee: 369356 km
21	00:48	LAST QUARTER MOON
21	04:52	Regulus 2.9°N of Moon
25	01	Venus 0.3°S of Mars
25	02:24	Spica 3.2°N of Moon
28	03:24	NEW MOON
30	07:53	Mars 1.6°N of Moon
30	10	Uranus at Opposition
30	13:20	Venus 0.9°N of Moon: Occn.
Dec 04	08:42	Moon at Apogee: 404663 km
06	05:22	FIRST QUARTER MOON
11	22	Mercury at Superior Conjunction
12	03:55	Pleiades 0.5°S of Moon
13	16:09	FULL MOON
14	20	Geminid Meteor Shower
16	02:21	Moon at Perigee: 364026 km
18	10:50	Regulus 3.1°N of Moon
20	09:11	LAST QUARTER MOON
22	02:43	Winter Solstice
22	08:13	Spica 3.3°N of Moon
23	04	Ursid Meteor Shower
25	18:01	Antares 0.4°S of Moon
27	20:12	NEW MOON
29	12:10	Mars 0.5°S of Moon: Occn.
31	01:54	Venus 3.1°S of Moon

Almanac of Astronomical Events for 2028

Date	GMT (h:m)	Event
Jan 01	03:52	Moon at Apogee: 405633 km
04	10	Quadrantid Meteor Shower
05	01:40	FIRST QUARTER MOON
05	13	Earth at Perihelion: 0.98331 AU
08	14:28	Pleiades 0.6°S of Moon
09	00	Mercury 0.7°S of Mars
12	04:03	FULL MOON
12	04:13	Partial Lunar Eclipse; mag=0.066
13	07:47	Moon at Perigee: 359046 km
14	19:13	Regulus 3.2°N of Moon
17	17	Mercury at Greatest Elong: 18.8°E
18	13:58	Spica 3.4°N of Moon
18	19:26	LAST QUARTER MOON
21	23:46	Antares 0.4°S of Moon
26	15:08	Annular Solar Eclipse; mag=0.921
26	15:12	NEW MOON
28	15:31	Moon at Apogee: 406387 km
Feb 02	09	Mercury at Inferior Conjunction
03	19:10	FIRST QUARTER MOON
04	23:48	Pleiades 0.6°S of Moon
10	15:04	FULL MOON
10	19:53	Moon at Perigee: 356678 km
11	06:05	Regulus 3.2°N of Moon
11	12	Mars at Perihelion: 1.38116 AU
14	21:42	Spica 3.4°N of Moon
17	08:08	LAST QUARTER MOON
18	05:37	Antares 0.4°S of Moon
22	23:55	Mercury 0.2°N of Moon: Occn.
24	16:26	Moon at Apogee: 406563 km
25	10:37	NEW MOON
27	16	Mercury at Greatest Elong: 26.9°W
28	18	Venus 3.4°N of Saturn
29	15:48	Venus 4.3°S of Moon
Mar 03	06:48	Pleiades 0.5°S of Moon
04	09:02	FIRST QUARTER MOON
09	17:30	Regulus 3.2°N of Moon
10	08:23	Moon at Perigee: 357599 km
11	01:06	FULL MOON
12	15	Jupiter at Opposition
13	07:43	Spica 3.3°N of Moon
16	13:06	Antares 0.6°S of Moon
17	23:23	LAST QUARTER MOON
20	02:17	Vernal Equinox
21	01	Mars in Conjunction with Sun
22	13	Venus at Greatest Elong: 46.1°E
22	23:24	Moon at Apogee: 406096 km
26	04:31	NEW MOON
26	10	Neptune in Conjunction with Sun
30	04:59	Venus 0.9°S of Moon: Occn.
30	12:17	Pleiades 0.3°S of Moon

Date	GMT (h:m)	Event
Apr 02	19:15	FIRST QUARTER MOON
03	17:50	Venus 0.2°S of Pleiades
06	03:15	Regulus 3.4°N of Moon
07	16:02	Moon at Perigee: 361361 km
09	10:27	FULL MOON
09	18:27	Spica 3.2°N of Moon
11	23	Mercury at Superior Conjunction
12	22:20	Antares 0.8°S of Moon
16	16:37	LAST QUARTER MOON
19	15:13	Moon at Apogee: 405100 km
20	12	Saturn in Conjunction with Sun
22	07	Lyrid Meteor Shower
24	19:47	NEW MOON
26	04:15	Mercury 2.9°S of Moon
26	18:05	Pleiades 0.1°S of Moon
28	00:20	Venus 1.7°N of Moon
May 01	01:33	Mercury 1.7°S of Pleiades
02	02:26	FIRST QUARTER MOON
03	10:26	Regulus 3.6°N of Moon
04	21	Eta-Aquarid Meteor Shower
05	10:31	Moon at Perigee: 366533 km
07	04:00	Spica 3.2°N of Moon
08	19:49	FULL MOON
09	06	Mercury at Greatest Elong: 21.4°E
10	08:04	Antares 1.0°S of Moon
16	10:43	LAST QUARTER MOON
17	09:55	Moon at Apogee: 404276 km
24	08:16	NEW MOON
30	16:01	Regulus 3.8°N of Moon
31	01	Uranus in Conjunction with Sun
31	06:18	Moon at Perigee: 369758 km
31	07:37	FIRST QUARTER MOON
Jun 01	10	Venus at Inferior Conjunction
02	03	Mercury at Inferior Conjunction
03	11:21	Spica 3.3°N of Moon
06	16:48	Antares 1.0°S of Moon
07	06:09	FULL MOON
13	23	Venus 2.0°S of Mars
14	04:46	Moon at Apogee: 404222 km
15	04:27	LAST QUARTER MOON
20	10:15	Pleiades 0.1°S of Moon
20	20:02	Summer Solstice
21	02:18	Mercury 1.4°N of Aldebaran
21	04:26	Mars 3.5°S of Moon
22	18:27	NEW MOON
26	04:14	Moon at Perigee: 366533 km
26	19	Mercury at Greatest Elong: 22.2°W
26	22:00	Regulus 4.0°N of Moon
29	12:11	FIRST QUARTER MOON
29	17	Mercury 3.3°S of Mars
30	17:02	Spica 3.5°N of Moon

Date	GMT (h:m)	Event
Jul 03	23	Earth at Aphelion: 1.01668 AU
03	23:48	Antares 0.9°S of Moon
06	18:11	FULL MOON
06	18:20	Partial Lunar Eclipse; mag=0.389
11	22:26	Moon at Apogee: 404945 km
13	01:06	Venus 1.0°N of Aldebaran
14	20:57	LAST QUARTER MOON
17	19:37	Pleiades 0.2°S of Moon
20	00:04	Mars 1.7°S of Moon
22	02:55	Total Solar Eclipse; mag=1.056
22	03:02	NEW MOON
23	22:14	Moon at Perigee: 361633 km
24	05:55	Regulus 4.0°N of Moon
25	13	Mercury at Superior Conjunction
27	22	Delta-Aquarid Meteor Shower
27	22:42	Spica 3.5°N of Moon
28	17:40	FIRST QUARTER MOON
31	05:27	Antares 0.9°S of Moon
Aug 05	08:10	FULL MOON
08	12:54	Moon at Apogee: 405894 km
10	17	Venus at Greatest Elong: 45.8°W
12	14	Perseid Meteor Shower
13	11:45	LAST QUARTER MOON
14	04:12	Pleiades 0.1°S of Moon
17	02:17	Venus 4.2°S of Moon
17	18:48	Mars 0.4°N of Moon: Occn.
20	10:44	NEW MOON
21	04:10	Moon at Perigee: 358074 km
22	04:46	Mercury 4.7°N of Moon
24	06:03	Spica 3.4°N of Moon
27	01:36	FIRST QUARTER MOON
27	11:09	Antares 1.1°S of Moon
27	22	Mercury 2.2°S of Jupiter
Sep 03	23:48	FULL MOON
04	20:03	Moon at Apogee: 406398 km
06	10	Mercury at Greatest Elong: 27.1°E
08	11	Venus 2.3°S of Mars
10	11:09	Pleiades 0.1°N of Moon
12	00:46	LAST QUARTER MOON
15	12:19	Mars 2.6°N of Moon
15	18:23	Venus 1.1°N of Moon: Occn.
17	02:49	Regulus 4.0°N of Moon
18	14:23	Moon at Perigee: 357048 km
18	18:24	NEW MOON
20	06:06	Mercury 1.2°N of Moon: Occn.
20	15:39	Spica 3.2°N of Moon
22	11:45	Autumnal Equinox
23	18:22	Antares 1.3°S of Moon
25	13:10	FIRST QUARTER MOON
30	02	Neptune at Opposition
30	13	Jupiter in Conjunction with Sun

Date	GMT (h:m)	Event
Oct 01	21:38	Moon at Apogee: 406246 km
02	06:44	Venus 0.1°S of Regulus
02	11	Mercury at Inferior Conjunction
03	16:25	FULL MOON
07	16:49	Pleiades 0.3°N of Moon
11	11:57	LAST QUARTER MOON
14	03:41	Mars 4.7°N of Moon
14	12:45	Regulus 4.2°N of Moon
17	00:41	Moon at Perigee: 359010 km
17	23	Mercury at Greatest Elong: 18.2°W
18	02:57	NEW MOON
21	03:41	Antares 1.5°S of Moon
21	07	Orionid Meteor Shower
22	14	Mercury 1.0°N of Jupiter
23	17:54	Mars 1.0°N of Regulus
25	04:53	FIRST QUARTER MOON
29	06:55	Moon at Apogee: 405560 km
30	17	Saturn at Opposition
Nov 02	09:17	FULL MOON
03	22:34	Pleiades 0.5°N of Moon
05	07	S Taurid Meteor Shower
09	21:26	LAST QUARTER MOON
10	01	Venus 0.6°N of Jupiter
10	20:20	Regulus 4.4°N of Moon
12	06	N Taurid Meteor Shower
14	05:53	Moon at Perigee: 363613 km
14	12:44	Spica 3.2°N of Moon
16	01:15	Venus 3.6°N of Spica
16	13:18	NEW MOON
17	13	Leonid Meteor Shower
20	01	Mercury at Superior Conjunction
24	00:14	FIRST QUARTER MOON
26	00:10	Moon at Apogee: 404732 km
Dec 01	05:45	Pleiades 0.5°N of Moon
02	01:40	FULL MOON
03	21	Uranus at Opposition
09	05:39	LAST QUARTER MOON
11	12:44	Moon at Perigee: 369056 km
11	20:36	Spica 3.3°N of Moon
14	02	Geminid Meteor Shower
16	02:06	NEW MOON
21	08:20	Winter Solstice
22	10	Ursid Meteor Shower
22	13:06	Jupiter 3.1°N of Spica
23	21:00	Moon at Apogee: 404409 km
23	21:45	FIRST QUARTER MOON
28	03	Jupiter at Aphelion: 5.45386 AU
28	14:35	Pleiades 0.4°N of Moon
31	01	Mercury at Greatest Elong: 19.6°E
31	16:48	FULL MOON
31	16:52	Total Lunar Eclipse; mag=1.246

Almanac of Astronomical Events for 2029

Date	GMT (h:m)	Event
Jan 02	19	Earth at Perihelion: 0.98329 AU
03	16	Quadrantid Meteor Shower
05	04:16	Moon at Perigee: 368918 km
07	13:26	LAST QUARTER MOON
08	02:16	Spica 3.4°N of Moon
11	07:10	Antares 1.6°S of Moon
13	07:32	Venus 0.8°N of Moon: Occn.
14	17:13	Partial Solar Eclipse; mag=0.871
14	17:24	NEW MOON
16	08	Mercury at Inferior Conjunction
19	22	Mars at Aphelion: 1.66609 AU
20	18:08	Moon at Apogee: 404884 km
22	19:23	FIRST QUARTER MOON
24	23:55	Pleiades 0.4°N of Moon
30	06:03	FULL MOON
Feb 01	12:21	Moon at Perigee: 363336 km
04	07:59	Spica 3.3°N of Moon
05	21:52	LAST QUARTER MOON
07	12:42	Antares 1.7°S of Moon
09	02	Mercury at Greatest Elong: 25.7°W
11	04:59	Mercury 1.0°S of Moon: Occn.
13	10:31	NEW MOON
17	11:59	Moon at Apogee: 405812 km
21	08:16	Pleiades 0.6°N of Moon
21	15:10	FIRST QUARTER MOON
28	17:10	FULL MOON
Mar 01	18:30	Moon at Perigee: 358626 km
03	15:56	Spica 3.1°N of Moon
06	18:29	Antares 1.9°S of Moon
07	07:52	LAST QUARTER MOON
15	04:19	NEW MOON
16	21:33	Moon at Apogee: 406528 km
20	08:01	Vernal Equinox
20	15:01	Pleiades 0.9°N of Moon
23	07:33	FIRST QUARTER MOON
23	20	Venus at Superior Conjunction
25	08	Mars at Opposition
26	16	Mercury at Superior Conjunction
28	23	Neptune in Conjunction with Sun
30	02:26	FULL MOON
30	05:40	Moon at Perigee: 356665 km
31	02:19	Spica 3.0°N of Moon

Date	GMT (h:m)	Event
Apr 01	16:11	Jupiter 3.4°N of Spica
03	02:24	Antares 2.1°S of Moon
05	19:51	LAST QUARTER MOON
12	03	Jupiter at Opposition
12	23:04	Moon at Apogee: 406670 km
13	15	Mercury 4.0°N of Saturn
13	21:40	NEW MOON
15	15:24	Mercury 1.9°S of Moon
16	20:47	Pleiades 1.0°N of Moon
21	13	Mercury at Greatest Elong: 20.1°E
21	19:50	FIRST QUARTER MOON
22	13	Lyrid Meteor Shower
27	13:23	Spica 3.0°N of Moon
27	16:24	Moon at Perigee: 357829 km
28	10:37	FULL MOON
30	12:25	Antares 2.3°S of Moon
May 01	03:48	Mercury 2.4°S of Pleiades
04	17	Saturn in Conjunction with Sun
05	03	Eta-Aquarid Meteor Shower
05	09:48	LAST QUARTER MOON
10	07:00	Moon at Apogee: 406125 km
12	23	Mercury at Inferior Conjunction
13	13:42	NEW MOON
21	04:16	FIRST QUARTER MOON
24	10	Mercury 1.3°S of Saturn
24	23:08	Spica 3.1°N of Moon
25	22:19	Moon at Perigee: 361586 km
27	18:37	FULL MOON
27	22:58	Antares 2.4°S of Moon
Jun 04	01:19	LAST QUARTER MOON
04	18	Uranus in Conjunction with Sun
06	22:09	Moon at Apogee: 405105 km
08	12	Mercury at Greatest Elong: 23.9°W
10	09:31	Pleiades 1.1°N of Moon
12	03:51	NEW MOON
12	04:05	Partial Solar Eclipse; mag=0.458
13	23:51	Venus 2.3°N of Moon
19	09:54	FIRST QUARTER MOON
20	08:00	Mercury 3.4°N of Aldebaran
21	01:48	Summer Solstice
21	06:32	Spica 3.1°N of Moon
22	15:35	Moon at Perigee: 366596 km
24	08:18	Antares 2.3°S of Moon
26	03:22	Total Lunar Eclipse; mag=1.844
26	03:22	FULL MOON

Date	GMT (h:m)	Event
Jul 03	17:58	LAST QUARTER MOON
04	16:05	Moon at Apogee: 404317 km
06	05	Earth at Aphelion: 1.01671 AU
07	17:23	Pleiades 1.1°N of Moon
09	20	Mercury at Superior Conjunction
11	15:36	Partial Solar Eclipse; mag=0.230
11	15:51	NEW MOON
18	11:25	Moon at Perigee: 369666 km
18	12:07	Spica 3.1°N of Moon
18	14:14	FIRST QUARTER MOON
19	16	Mars 1.6°S of Jupiter
21	09:14	Venus 1.0°N of Regulus
21	15:31	Antares 2.4°S of Moon
25	13:36	FULL MOON
28	01:29	Mars 1.4°N of Spica
28	04	Delta-Aquarid Meteor Shower
31	21:28	Mercury 0.4°N of Regulus
Aug 01	10:42	Moon at Apogee: 404310 km
02	11:15	LAST QUARTER MOON
04	01:43	Pleiades 1.2°N of Moon
10	01:56	NEW MOON
12	04:13	Mercury 4.6°N of Moon
12	20	Perseid Meteor Shower
13	09:55	Moon at Perigee: 366368 km
14	17:41	Spica 2.9°N of Moon
15	12:04	Mars 3.7°N of Moon
16	18:55	FIRST QUARTER MOON
17	21:04	Antares 2.5°S of Moon
19	21	Mercury at Greatest Elong: 27.4°E
22	13:20	Jupiter 2.9°N of Spica
24	01:51	FULL MOON
29	04:45	Moon at Apogee: 405066 km
31	09:39	Pleiades 1.4°N of Moon
Sep 01	04:33	LAST QUARTER MOON
05	04:33	Venus 1.4°N of Spica
07	12	Venus 1.7°S of Jupiter
08	10:44	NEW MOON
10	04:27	Moon at Perigee: 361349 km
11	01:06	Spica 2.7°N of Moon
11	13:29	Venus 3.6°N of Moon
12	23:54	Mars 2.1°N of Moon
14	02:41	Antares 2.7°S of Moon
15	01:29	FIRST QUARTER MOON
15	21	Mercury at Inferior Conjunction
22	16:29	FULL MOON
22	17:37	Autumnal Equinox
25	19:40	Moon at Apogee: 406002 km
27	16:36	Pleiades 1.6°N of Moon
30	20:57	LAST QUARTER MOON

Date	GMT (h:m)	Event
Oct 01	15	Mercury at Greatest Elong: 17.9°W
02	14	Neptune at Opposition
07	08:09	Mars 3.3°N of Antares
07	19:14	NEW MOON
08	11:26	Moon at Perigee: 357740 km
11	01:34	Venus 0.9°S of Moon: Occn.
11	10:16	Antares 3.0°S of Moon
11	16:13	Mars 0.1°N of Moon: Occn.
14	11:09	FIRST QUARTER MOON
16	08:50	Venus 1.3°N of Antares
21	13	Orionid Meteor Shower
22	09:28	FULL MOON
23	02:02	Moon at Apogee: 406433 km
24	22:44	Pleiades 1.8°N of Moon
27	12	Venus at Greatest Elong: 47.0°E
30	11:32	LAST QUARTER MOON
30	23	Mercury at Superior Conjunction
30	23	Jupiter in Conjunction with Sun
Nov 04	22:14	Spica 2.7°N of Moon
05	13	S Taurid Meteor Shower
05	23:09	Moon at Perigee: 356900 km
06	04:24	NEW MOON
07	20:23	Antares 3.1°S of Moon
09	12:30	Venus 4.6°S of Moon
09	13:20	Mars 2.1°S of Moon
13	00:35	FIRST QUARTER MOON
13	15	Saturn at Opposition
17	19	Leonid Meteor Shower
19	02:53	Moon at Apogee: 406241 km
21	04:03	FULL MOON
21	04:46	Pleiades 1.8°N of Moon
28	23:48	LAST QUARTER MOON
29	12	Venus 1.9°S of Mars
Dec 02	08:35	Spica 2.7°N of Moon
03	19:14	Jupiter 4.5°N of Moon
04	10:38	Moon at Perigee: 359274 km
05	14:52	NEW MOON
05	15:03	Partial Solar Eclipse; mag=0.891
07	00:20	Mercury 3.2°S of Moon
08	08:11	Venus 4.6°S of Moon
08	09	Uranus at Opposition
08	14:50	Mars 4.1°S of Moon
12	17:49	FIRST QUARTER MOON
14	03	Mercury at Greatest Elong: 20.6°E
14	08	Geminid Meteor Shower
16	13:58	Moon at Apogee: 405520 km
18	11:32	Pleiades 1.8°N of Moon
20	22:42	Total Lunar Eclipse; mag=1.117
20	22:46	FULL MOON
21	14:14	Winter Solstice
22	16	Ursid Meteor Shower
28	09:49	LAST QUARTER MOON
29	13	Mars at Perihelion: 1.38140 AU
29	16:17	Spica 2.7°N of Moon
31	12	Mercury at Inferior Conjunction
31	12:42	Jupiter 4.1°N of Moon

Almanac of Astronomical Events for 2030

Date	GMT (h:m)	Event
Jan 01	15:33	Moon at Perigee: 364250 km
01	17:49	Antares 3.1°S of Moon
03	11	Earth at Perihelion: 0.98334 AU
03	22	Quadrantid Meteor Shower
04	02:49	NEW MOON
06	13	Venus at Inferior Conjunction
11	14:06	FIRST QUARTER MOON
13	08:51	Moon at Apogee: 404647 km
14	19:12	Pleiades 1.8°N of Moon
19	15:54	FULL MOON
22	10	Mercury at Greatest Elong: 24.4°W
25	21:46	Spica 2.6°N of Moon
26	18:14	LAST QUARTER MOON
28	02:02	Jupiter 3.7°N of Moon
28	16:03	Moon at Perigee: 369541 km
29	01:14	Antares 3.2°S of Moon
31	22:04	Mercury 2.4°S of Moon
Feb 02	16:07	NEW MOON
10	06:06	Moon at Apogee: 404292 km
10	11:49	FIRST QUARTER MOON
11	03:22	Pleiades 2.0°N of Moon
18	06:20	FULL MOON
22	03:28	Spica 2.4°N of Moon
22	10:01	Moon at Perigee: 368405 km
24	11:19	Jupiter 3.3°N of Moon
25	01:58	LAST QUARTER MOON
25	06:37	Antares 3.5°S of Moon
28	13:29	Venus 1.4°N of Moon
Mar 04	06:35	NEW MOON
09	23	Mercury at Superior Conjunction
10	02:23	Moon at Apogee: 404748 km
10	11:18	Pleiades 2.3°N of Moon
12	08:48	FIRST QUARTER MOON
17	23	Venus at Greatest Elong: 46.6°W
19	17:56	FULL MOON
20	13:51	Vernal Equinox
21	11:29	Spica 2.2°N of Moon
21	22:00	Moon at Perigee: 362974 km
23	18:08	Jupiter 3.1°N of Moon
24	12:31	Antares 3.7°S of Moon
26	09:51	LAST QUARTER MOON
30	01:16	Venus 3.8°S of Moon
31	11	Neptune in Conjunction with Sun

Date	GMT (h:m)	Event
Apr 02	22:02	NEW MOON
04	08	Mercury at Greatest Elong: 19.1°E
04	15:32	Mercury 0.9°S of Moon: Occn.
06	07:31	Saturn 4.1°S of Moon
06	18:29	Pleiades 2.5°N of Moon
06	18:47	Moon at Apogee: 405659 km
11	02:57	FIRST QUARTER MOON
17	21:45	Spica 2.2°N of Moon
18	03:20	FULL MOON
19	03:44	Moon at Perigee: 358706 km
20	00:03	Jupiter 3.1°N of Moon
20	20:44	Antares 3.9°S of Moon
22	20	Lyrid Meteor Shower
23	11	Mercury at Inferior Conjunction
24	18:39	LAST QUARTER MOON
May 02	14:12	NEW MOON
04	00:53	Pleiades 2.6°N of Moon
04	03:39	Moon at Apogee: 406364 km
05	09	Eta-Aquarid Meteor Shower
10	17:11	FIRST QUARTER MOON
13	10	Jupiter at Opposition
15	08:36	Spica 2.2°N of Moon
17	05:49	Jupiter 3.4°N of Moon
17	11:19	FULL MOON
17	13:45	Moon at Perigee: 357018 km
18	07:06	Antares 4.0°S of Moon
19	06	Saturn in Conjunction with Sun
21	02	Mercury at Greatest Elong: 25.6°W
24	04:57	LAST QUARTER MOON
25	10	Mars in Conjunction with Sun
31	06:14	Moon at Apogee: 406469 km
Jun 01	06:21	NEW MOON
01	06:28	Annular Solar Eclipse; mag=0.944
08	17	Mercury 0.3°N of Saturn
09	03:36	FIRST QUARTER MOON
09	13	Uranus in Conjunction with Sun
11	18:08	Spica 2.3°N of Moon
13	11:42	Jupiter 3.6°N of Moon
14	17:58	Antares 4.0°S of Moon
14	23:37	Moon at Perigee: 358183 km
15	18:33	Partial Lunar Eclipse; mag=0.503
15	18:41	FULL MOON
21	07:31	Summer Solstice
22	17:19	LAST QUARTER MOON
24	07	Mercury at Superior Conjunction
24	21	Venus 0.3°N of Saturn
27	13:09	Pleiades 2.6°N of Moon
27	14:14	Moon at Apogee: 405881 km
27	22:12	Saturn 3.1°S of Moon
28	05:11	Venus 2.3°S of Moon
30	21:34	NEW MOON

Date	GMT (h:m)	Event
Jul 04	13	Earth at Aphelion: 1.01672 AU
08	11:02	FIRST QUARTER MOON
09	01:21	Spica 2.2°N of Moon
10	17:59	Jupiter 3.5°N of Moon
12	03:31	Antares 4.0°S of Moon
13	05:12	Moon at Perigee: 361792 km
15	02:12	FULL MOON
22	08:07	LAST QUARTER MOON
24	20:03	Pleiades 2.7°N of Moon
25	04:55	Moon at Apogee: 404891 km
25	10:44	Saturn 2.7°S of Moon
26	09:59	Mercury 0.6°S of Regulus
28	11	Delta-Aquarid Meteor Shower
28	13:10	Venus 2.3°N of Moon
28	23:18	Mars 3.7°N of Moon
30	11:11	NEW MOON
Aug 01	20:50	Mercury 4.0°N of Moon
02	05	Mercury at Greatest Elong: 27.3°E
05	06:48	Spica 2.0°N of Moon
05	21	Venus 0.7°S of Mars
06	16:43	FIRST QUARTER MOON
07	01:26	Jupiter 3.2°N of Moon
08	10:46	Antares 4.1°S of Moon
09	22:50	Moon at Perigee: 366742 km
13	03	Perseid Meteor Shower
13	10:44	FULL MOON
21	01:15	LAST QUARTER MOON
21	03:40	Pleiades 2.9°N of Moon
21	22:15	Saturn 2.4°S of Moon
21	22:49	Moon at Apogee: 404174 km
28	23:07	NEW MOON
29	20	Mercury at Inferior Conjunction
Sep 01	12:19	Spica 1.8°N of Moon
03	11:11	Jupiter 2.6°N of Moon
04	16:16	Antares 4.4°S of Moon
04	17:06	Moon at Perigee: 369884 km
04	21:56	FIRST QUARTER MOON
10	04:17	Mercury 1.8°S of Regulus
11	21:18	FULL MOON
15	06	Mercury at Greatest Elong: 17.9°W
17	11:36	Pleiades 3.2°N of Moon
18	07:45	Saturn 2.1°S of Moon
18	18:09	Moon at Apogee: 404258 km
19	19:56	LAST QUARTER MOON
22	23:27	Autumnal Equinox
25	07:11	Saturn 3.2°N of Aldebaran
27	09:55	NEW MOON
28	19:44	Spica 1.6°N of Moon
30	15:39	Moon at Perigee: 366348 km

Date	GMT (h:m)	Event
Oct 01	00:24	Jupiter 1.9°N of Moon
02	03:17	Mars 0.8°N of Regulus
04	03:56	FIRST QUARTER MOON
05	03	Neptune at Opposition
11	10:47	FULL MOON
11	23	Mercury at Superior Conjunction
14	19:18	Pleiades 3.4°N of Moon
15	14:21	Saturn 2.1°S of Moon
16	13:21	Moon at Apogee: 405078 km
19	14:50	LAST QUARTER MOON
20	11	Venus at Superior Conjunction
21	19	Orionid Meteor Shower
26	20:17	NEW MOON
28	12:07	Moon at Perigee: 361122 km
28	17:29	Jupiter 1.2°N of Moon: Occn.
Nov 02	11:56	FIRST QUARTER MOON
05	19	S Taurid Meteor Shower
09	10	Mercury 2.6°S of Jupiter
10	03:30	FULL MOON
11	02:18	Pleiades 3.4°N of Moon
11	18:03	Saturn 2.2°S of Moon
12	19	N Taurid Meteor Shower
13	05:00	Moon at Apogee: 405993 km
13	20:29	Mercury 2.3°N of Antares
18	01	Leonid Meteor Shower
18	08:32	LAST QUARTER MOON
22	16:31	Spica 1.7°N of Moon
25	06:46	NEW MOON
25	06:50	Total Solar Eclipse; mag=1.047
25	21:07	Moon at Perigee: 357521 km
26	19:19	Mercury 4.1°S of Moon
26	23	Mercury at Greatest Elong: 21.8°E
27	16	Saturn at Opposition
30	13	Jupiter in Conjunction with Sun
Dec 01	22:57	FIRST QUARTER MOON
08	00	Mars at Aphelion: 1.66601 AU
08	08:35	Pleiades 3.4°N of Moon
08	20:06	Saturn 2.3°S of Moon
09	22:28	Pen. Lunar Eclipse; mag=0.942
09	22:40	FULL MOON
10	10:03	Moon at Apogee: 406371 km
12	21	Uranus at Opposition
14	14	Geminid Meteor Shower
15	19	Mercury at Inferior Conjunction
18	00:01	LAST QUARTER MOON
20	02:22	Spica 1.6°N of Moon
21	20:09	Winter Solstice
22	23	Ursid Meteor Shower
23	10:51	Jupiter 0.0°S of Moon: Occn.
23	15:08	Mercury 2.2°N of Moon
24	10:09	Moon at Perigee: 356926 km
24	17:32	NEW MOON
25	22:23	Venus 5.0°S of Moon
28	10	Mercury 2.6°N of Jupiter
31	13:36	FIRST QUARTER MOON

Almanac of Astronomical Events for 2031

Date	GMT (h:m)	Event
Jan 03	14:09	Mars 3.6°N of Spica
04	04	Quadrantid Meteor Shower
04	14:42	Pleiades 3.5°N of Moon
04	20	Mercury at Greatest Elong: 22.9°W
04	21	Earth at Perihelion: 0.98326 AU
04	22:42	Saturn 2.4°S of Moon
06	11:00	Moon at Apogee: 406169 km
08	18:26	FULL MOON
16	09:36	Spica 1.5°N of Moon
16	12:47	LAST QUARTER MOON
20	06:34	Jupiter 0.6°S of Moon: Occn.
21	21:41	Moon at Perigee: 359637 km
21	21:53	Mercury 3.7°S of Moon
23	04:31	NEW MOON
30	07:43	FIRST QUARTER MOON
31	21:26	Pleiades 3.7°N of Moon
Feb 01	03:57	Saturn 2.2°S of Moon
02	23:45	Moon at Apogee: 405416 km
07	12:46	FULL MOON
12	14:57	Spica 1.2°N of Moon
14	01:19	Mars 3.7°N of Moon
14	22:50	LAST QUARTER MOON
16	22:35	Jupiter 1.3°S of Moon: Occn.
19	00:35	Moon at Perigee: 364745 km
20	13	Mercury at Superior Conjunction
21	15:49	NEW MOON
28	05:14	Pleiades 3.9°N of Moon
28	12:56	Saturn 1.7°S of Moon
Mar 01	04:02	FIRST QUARTER MOON
02	18:57	Moon at Apogee: 404533 km
09	04:30	FULL MOON
11	20:44	Spica 1.0°N of Moon
13	19:27	Mars 2.7°N of Moon
16	06:36	LAST QUARTER MOON
16	10:06	Jupiter 1.8°S of Moon
17	18:58	Moon at Perigee: 369634 km
18	12	Mercury at Greatest Elong: 18.5°E
20	19:41	Vernal Equinox
23	03:49	NEW MOON
24	10:55	Mercury 0.1°S of Moon: Occn.
26	04:40	Venus 0.9°S of Moon: Occn.
28	01:01	Saturn 1.2°S of Moon
30	15:24	Moon at Apogee: 404227 km
31	00:32	FIRST QUARTER MOON

Date	GMT (h:m)	Event
Apr 02	23	Neptune in Conjunction with Sun
04	20	Mercury at Inferior Conjunction
07	17:21	FULL MOON
08	04:35	Spica 0.9°N of Moon
10	02:38	Mars 2.1°N of Moon
10	18:05	Venus 2.4°S of Pleiades
11	19:13	Moon at Perigee: 367996 km
12	17:56	Jupiter 2.1°S of Moon
14	12:58	LAST QUARTER MOON
16	03	Venus 3.4°N of Saturn
21	16:57	NEW MOON
23	02	Lyrid Meteor Shower
24	14:42	Saturn 0.7°S of Moon: Occn.
25	08:28	Venus 3.9°N of Moon
27	10:30	Moon at Apogee: 404758 km
29	19:19	FIRST QUARTER MOON
May 02	20	Mercury at Greatest Elong: 26.9°W
04	11	Mars at Opposition
05	14:17	Spica 1.0°N of Moon
05	15	Eta-Aquarid Meteor Shower
06	22:45	Mars 1.9°N of Moon
07	03:40	FULL MOON
07	03:51	Pen. Lunar Eclipse; mag=0.881
09	07:28	Moon at Perigee: 362851 km
09	23:39	Jupiter 2.1°S of Moon
13	19:07	LAST QUARTER MOON
14	06:50	Saturn 3.9°N of Aldebaran
19	14:37	Mercury 4.4°S of Moon
21	07:15	Annular Solar Eclipse; mag=0.959
21	07:17	NEW MOON
25	02:08	Moon at Apogee: 405723 km
29	09:28	Venus 3.9°S of Pollux
29	11:20	FIRST QUARTER MOON
Jun 02	00:22	Spica 1.0°N of Moon
02	03	Venus at Greatest Elong: 45.4°E
02	19:58	Mars 1.4°N of Moon
03	03	Saturn in Conjunction with Sun
05	11:44	Pen. Lunar Eclipse; mag=0.129
05	11:58	FULL MOON
06	04:42	Jupiter 1.9°S of Moon
06	12:11	Moon at Perigee: 358744 km
08	19	Mercury at Superior Conjunction
12	02:21	LAST QUARTER MOON
14	08	Uranus in Conjunction with Sun
15	07	Jupiter at Opposition
19	22:25	NEW MOON
21	11:25	Moon at Apogee: 406452 km
21	13:17	Summer Solstice
28	00:19	FIRST QUARTER MOON
29	09:14	Spica 0.9°N of Moon
30	06:41	Mars 0.4°N of Moon: Occn.

27

Date	GMT (h:m)	Event
Jul 03	10:02	Jupiter 1.7°S of Moon
04	19:01	FULL MOON
04	21:14	Moon at Perigee: 357008 km
06	06	Earth at Aphelion: 1.01669 AU
11	11:50	LAST QUARTER MOON
15	07	Mercury at Greatest Elong: 26.6°E
16	06:23	Saturn 0.4°N of Moon: Occn.
18	14:28	Moon at Apogee: 406535 km
19	13:40	NEW MOON
21	23:10	Mercury 3.3°N of Moon
22	02:15	Venus 1.6°N of Moon
25	18	Mercury 1.6°N of Venus
26	16:05	Spica 0.6°N of Moon
27	10:35	FIRST QUARTER MOON
28	06:06	Mars 1.1°S of Moon: Occn.
28	17	Delta-Aquarid Meteor Shower
30	16:09	Jupiter 1.7°S of Moon
Aug 02	06:47	Moon at Perigee: 358060 km
03	01:45	FULL MOON
10	00:24	LAST QUARTER MOON
11	03	Venus at Inferior Conjunction
12	04	Mercury at Inferior Conjunction
12	18:11	Saturn 0.7°N of Moon: Occn.
13	09	Perseid Meteor Shower
14	21:37	Moon at Apogee: 405946 km
18	04:32	NEW MOON
22	21:30	Spica 0.4°N of Moon
25	13:38	Mars 2.8°S of Moon
25	18:40	FIRST QUARTER MOON
26	23:28	Jupiter 2.0°S of Moon
29	17	Mercury at Greatest Elong: 18.2°W
30	12:53	Moon at Perigee: 361684 km
Sep 01	09:20	FULL MOON
08	05:21	Aldebaran 3.8°S of Moon
08	16:14	LAST QUARTER MOON
09	05:03	Saturn 1.1°N of Moon: Occn.
10	11:09	Mars 2.5°N of Antares
11	11:54	Moon at Apogee: 405006 km
13	10:17	Venus 0.4°S of Moon: Occn.
16	18:47	NEW MOON
19	03:05	Spica 0.2°N of Moon
23	02:35	Mars 4.4°S of Moon
23	05:15	Autumnal Equinox
23	08:42	Jupiter 2.5°S of Moon
24	01:20	FIRST QUARTER MOON
24	01	Mercury at Superior Conjunction
27	07:09	Moon at Perigee: 366841 km
29	01	Mars 2.2°S of Jupiter
30	18:58	FULL MOON

Date	GMT (h:m)	Event
Oct 05	13:48	Aldebaran 3.6°S of Moon
06	14:25	Saturn 1.3°N of Moon
07	15	Neptune at Opposition
08	10:50	LAST QUARTER MOON
09	06:23	Moon at Apogee: 404365 km
09	10:13	Venus 2.1°S of Regulus
12	13:45	Venus 4.0°N of Moon
16	08:21	NEW MOON
17	14:11	Mercury 0.2°S of Moon: Occn.
20	20:53	Jupiter 3.0°S of Moon
21	13	Venus at Greatest Elong: 46.4°W
22	01	Orionid Meteor Shower
22	20:06	Moon at Perigee: 370022 km
23	07:36	FIRST QUARTER MOON
30	07:33	FULL MOON
30	07:45	Pen. Lunar Eclipse; mag=0.716
Nov 01	22:35	Aldebaran 3.6°S of Moon
02	21:32	Saturn 1.3°N of Moon
06	01	S Taurid Meteor Shower
06	02:45	Moon at Apogee: 404522 km
07	07:02	LAST QUARTER MOON
09	10:11	Mercury 1.8°N of Antares
09	15	Mercury at Greatest Elong: 23.1°E
12	19:40	Spica 0.2°N of Moon
13	01	N Taurid Meteor Shower
14	21:06	Hybrid Solar Eclipse; mag=1.011
14	21:10	NEW MOON
16	08	Mars at Perihelion: 1.38145 AU
16	10:48	Mercury 4.7°S of Moon
17	12:51	Jupiter 3.4°S of Moon
17	22:07	Moon at Perigee: 365930 km
18	07	Leonid Meteor Shower
21	14:45	FIRST QUARTER MOON
28	23:18	FULL MOON
29	06:36	Aldebaran 3.6°S of Moon
30	02:00	Saturn 1.1°N of Moon: Occn.
30	03	Mercury at Inferior Conjunction
Dec 03	22:35	Moon at Apogee: 405355 km
07	03:20	LAST QUARTER MOON
10	05:42	Spica 0.2°N of Moon
11	09:18	Venus 3.3°N of Moon
11	19	Saturn at Opposition
12	22:17	Mercury 1.6°N of Moon
14	09:06	NEW MOON
14	20	Geminid Meteor Shower
15	21:31	Moon at Perigee: 360489 km
17	11	Uranus at Opposition
18	10	Mercury at Greatest Elong: 21.5°W
21	00:00	FIRST QUARTER MOON
22	01:56	Winter Solstice
23	05	Ursid Meteor Shower
26	13:13	Aldebaran 3.6°S of Moon
27	04:28	Saturn 1.0°N of Moon: Occn.
28	17:33	FULL MOON
31	13:16	Moon at Apogee: 406203 km

Almanac of Astronomical Events for 2032

Date	GMT (h:m)	Event
Jan 01	06	Jupiter in Conjunction with Sun
03	05	Earth at Perihelion: 0.98325 AU
04	10	Quadrantid Meteor Shower
05	22:04	LAST QUARTER MOON
06	14:33	Spica 0.0°S of Moon
10	07:47	Venus 0.8°S of Moon: Occn.
12	20:07	NEW MOON
13	07:55	Moon at Perigee: 357088 km
19	12:14	FIRST QUARTER MOON
22	18:54	Aldebaran 3.4°S of Moon
23	06:57	Saturn 1.1°N of Moon: Occn.
27	12:52	FULL MOON
27	16:13	Moon at Apogee: 406533 km
Feb 02	07	Mercury at Superior Conjunction
02	21:18	Spica 0.3°S of Moon
04	13:49	LAST QUARTER MOON
07	01	Venus 0.3°N of Jupiter
09	03:46	Jupiter 4.7°S of Moon
09	07:35	Venus 4.5°S of Moon
10	20:46	Moon at Perigee: 356948 km
11	06:24	NEW MOON
14	06:38	Mars 3.6°S of Moon
18	03:29	FIRST QUARTER MOON
19	01:05	Aldebaran 3.2°S of Moon
19	12:01	Saturn 1.4°N of Moon
23	18:41	Moon at Apogee: 406290 km
26	07:43	FULL MOON
29	22	Mercury at Greatest Elong: 18.2°E
Mar 01	02:47	Spica 0.6°S of Moon
05	01:47	LAST QUARTER MOON
10	06:54	Moon at Perigee: 360020 km
11	16:25	NEW MOON
14	03:55	Mars 1.3°S of Moon
17	01	Mercury at Inferior Conjunction
17	08:53	Aldebaran 2.9°S of Moon
17	21:15	Saturn 1.8°N of Moon
18	20:56	FIRST QUARTER MOON
20	01:23	Vernal Equinox
22	08:27	Moon at Apogee: 405454 km
27	00:46	FULL MOON
27	11	Mercury 2.7°N of Venus
28	08:45	Spica 0.6°S of Moon

Date	GMT (h:m)	Event
Apr 03	10:10	LAST QUARTER MOON
04	12	Neptune in Conjunction with Sun
07	06:55	Moon at Perigee: 365117 km
10	02:39	NEW MOON
12	01:18	Mars 1.1°N of Moon: Occn.
13	18:03	Aldebaran 2.8°S of Moon
13	21	Mercury at Greatest Elong: 27.7°W
14	10:08	Saturn 2.2°N of Moon
17	15:24	FIRST QUARTER MOON
19	03:02	Moon at Apogee: 404547 km
22	08	Lyrid Meteor Shower
24	16:11	Spica 0.6°S of Moon
25	15:10	FULL MOON
25	15:14	Total Lunar Eclipse; mag=1.191
28	20:29	Mars 3.5°S of Pleiades
May 02	16:02	LAST QUARTER MOON
03	20:49	Moon at Perigee: 369512 km
04	21	Eta-Aquarid Meteor Shower
08	08:42	Mercury 2.7°S of Moon
09	13:25	Annular Solar Eclipse; mag=0.996
09	13:36	NEW MOON
10	22:56	Mars 3.1°N of Moon
11	03:14	Aldebaran 2.8°S of Moon
12	00:51	Saturn 2.4°N of Moon
16	22:18	Moon at Apogee: 404274 km
17	09:43	FIRST QUARTER MOON
22	00:53	Spica 0.7°S of Moon
23	06	Mercury at Superior Conjunction
25	02:37	FULL MOON
29	02:49	Moon at Perigee: 367645 km
31	20:51	LAST QUARTER MOON
Jun 02	09	Venus at Superior Conjunction
08	01:32	NEW MOON
13	16:28	Moon at Apogee: 404843 km
16	03:00	FIRST QUARTER MOON
17	05	Saturn in Conjunction with Sun
18	05	Uranus in Conjunction with Sun
18	09:48	Spica 0.8°S of Moon
20	19:09	Summer Solstice
23	11:32	FULL MOON
25	14:55	Moon at Perigee: 362652 km
26	02	Mercury at Greatest Elong: 25.4°E
30	02:12	LAST QUARTER MOON

Date	GMT (h:m)	Event
Jul 04	17:33	Aldebaran 2.8°S of Moon
05	12	Earth at Aphelion: 1.01675 AU
06	05:21	Saturn 2.9°N of Moon
07	14:41	NEW MOON
09	10:33	Mercury 2.5°N of Moon
11	05	Mars in Conjunction with Sun
11	07:50	Moon at Apogee: 405826 km
15	17:51	Spica 1.0°S of Moon
15	18:32	FIRST QUARTER MOON
19	06	Jupiter at Opposition
22	18:51	FULL MOON
23	18	Mercury at Inferior Conjunction
23	18:45	Moon at Perigee: 358602 km
27	23	Delta-Aquarid Meteor Shower
29	09:25	LAST QUARTER MOON
31	23:01	Aldebaran 2.6°S of Moon
Aug 02	17:42	Saturn 3.2°N of Moon
04	09:57	Venus 1.0°N of Regulus
04	19:04	Mercury 1.8°N of Moon
06	05:11	NEW MOON
07	17:16	Moon at Apogee: 406532 km
11	21	Mercury at Greatest Elong: 18.9°W
12	00:29	Spica 1.3°S of Moon
12	15	Perseid Meteor Shower
14	07:51	FIRST QUARTER MOON
21	01:47	FULL MOON
21	03:52	Moon at Perigee: 356879 km
27	19:33	LAST QUARTER MOON
28	04:59	Aldebaran 2.3°S of Moon
30	04:50	Saturn 3.6°N of Moon
Sep 03	19:58	Moon at Apogee: 406561 km
04	20:57	NEW MOON
05	23	Mercury at Superior Conjunction
07	08:32	Venus 2.6°N of Moon
08	06:09	Spica 1.5°S of Moon
12	00:08	Mars 0.7°N of Regulus
12	18:49	FIRST QUARTER MOON
17	14:43	Venus 2.2°N of Spica
18	14:05	Moon at Perigee: 358032 km
19	09:30	FULL MOON
22	11:11	Autumnal Equinox
24	12:40	Aldebaran 2.2°S of Moon
26	09:12	LAST QUARTER MOON
26	14:58	Saturn 3.8°N of Moon
29	10:26	Mercury 1.3°N of Spica

Date	GMT (h:m)	Event
Oct 01	03:03	Moon at Apogee: 405951 km
04	13:26	NEW MOON
06	07:43	Mercury 1.8°S of Moon
07	10:45	Venus 2.3°S of Moon
09	04	Neptune at Opposition
12	03:48	FIRST QUARTER MOON
16	21:27	Moon at Perigee: 361921 km
18	18:58	FULL MOON
18	19:02	Total Lunar Eclipse; mag=1.103
21	07	Orionid Meteor Shower
21	22:10	Aldebaran 2.1°S of Moon
22	04	Mercury at Greatest Elong: 24.4°E
24	00:05	Saturn 3.9°N of Moon
24	23	Mars at Aphelion: 1.66598 AU
25	12:39	Venus 3.0°N of Antares
26	02:29	LAST QUARTER MOON
28	18:22	Moon at Apogee: 405007 km
31	01:27	Mars 4.1°N of Moon
Nov 01	18:58	Spica 1.5°S of Moon
03	05:33	Partial Solar Eclipse; mag=0.855
03	05:45	NEW MOON
05	08	S Taurid Meteor Shower
10	11:33	FIRST QUARTER MOON
13	09	Mercury at Inferior Conjunction
13	15:20	Moon at Perigee: 367372 km
17	06:42	FULL MOON
17	13	Leonid Meteor Shower
18	08:14	Aldebaran 2.1°S of Moon
20	07:39	Saturn 3.8°N of Moon
24	22:48	LAST QUARTER MOON
25	14:16	Moon at Apogee: 404371 km
28	16	Saturn at Perihelion: 9.01492 AU
28	19:33	Mars 2.4°N of Moon
29	03:15	Spica 1.5°S of Moon
30	08	Mercury at Greatest Elong: 20.2°W
Dec 01	05:34	Mercury 0.6°N of Moon: Occn.
02	20:53	NEW MOON
07	15:03	Mars 3.0°N of Spica
08	07	Venus 1.8°S of Jupiter
08	19:18	Moon at Perigee: 370105 km
09	19:09	FIRST QUARTER MOON
14	02	Geminid Meteor Shower
15	17:05	Aldebaran 2.1°S of Moon
16	20:49	FULL MOON
17	13:04	Saturn 3.6°N of Moon
21	00	Uranus at Opposition
21	07:57	Winter Solstice
22	11	Ursid Meteor Shower
23	11:34	Moon at Apogee: 404516 km
24	20:39	LAST QUARTER MOON
24	23	Saturn at Opposition
26	11:57	Spica 1.7°S of Moon
27	13:11	Mars 0.5°N of Moon: Occn.

Almanac of Astronomical Events for 2033

Date	GMT (h:m)	Event
Jan 01	10:17	NEW MOON
03	17	Quadrantid Meteor Shower
04	05:24	Moon at Perigee: 365354 km
04	12	Earth at Perihelion: 0.98330 AU
07	19	Venus at Greatest Elong: 47.2°E
08	03:34	FIRST QUARTER MOON
11	23:46	Aldebaran 2.0°S of Moon
13	00	Mercury at Superior Conjunction
13	16:25	Saturn 3.6°N of Moon
15	13:07	FULL MOON
20	07:04	Moon at Apogee: 405294 km
22	19:55	Spica 2.0°S of Moon
23	17:46	LAST QUARTER MOON
25	05:50	Mars 1.4°S of Moon
30	22:00	NEW MOON
Feb 01	07:27	Moon at Perigee: 360085 km
02	20	Jupiter in Conjunction with Sun
03	01:23	Venus 0.3°N of Moon: Occn.
06	13:34	FIRST QUARTER MOON
08	05:09	Aldebaran 1.8°S of Moon
09	19:16	Saturn 3.8°N of Moon
12	10	Mercury at Greatest Elong: 18.2°E
14	07:04	FULL MOON
16	19:57	Moon at Apogee: 406072 km
19	02:43	Spica 2.3°S of Moon
22	11:53	LAST QUARTER MOON
22	20:19	Mars 3.1°S of Moon
27	23	Mercury at Inferior Conjunction
Mar 01	08:23	NEW MOON
01	18:17	Moon at Perigee: 357180 km
07	11:22	Aldebaran 1.6°S of Moon
08	01:27	FIRST QUARTER MOON
09	00:25	Saturn 4.1°N of Moon
15	22:00	Moon at Apogee: 406368 km
16	01:37	FULL MOON
18	08:45	Spica 2.4°S of Moon
20	07:23	Vernal Equinox
20	16	Venus at Inferior Conjunction
23	06:59	Mars 4.4°S of Moon
24	01:50	LAST QUARTER MOON
27	02	Mercury at Greatest Elong: 27.8°W
30	06:09	Moon at Perigee: 357444 km
30	17:52	NEW MOON
30	18:01	Total Solar Eclipse; mag=1.046

Date	GMT (h:m)	Event
Apr 03	19:47	Aldebaran 1.5°S of Moon
05	09:44	Saturn 4.3°N of Moon
06	15:14	FIRST QUARTER MOON
07	00	Neptune in Conjunction with Sun
12	02:26	Moon at Apogee: 406060 km
14	14:53	Spica 2.4°S of Moon
14	19:13	Total Lunar Eclipse; mag=1.094
14	19:17	FULL MOON
22	11:42	LAST QUARTER MOON
22	14	Lyrid Meteor Shower
26	06:34	Venus 0.5°S of Moon: Occn.
27	14:44	Moon at Perigee: 360572 km
29	02:46	NEW MOON
May 01	05:52	Aldebaran 1.5°S of Moon
02	22:56	Saturn 4.4°N of Moon
05	03	Eta-Aquarid Meteor Shower
06	06:45	FIRST QUARTER MOON
07	15	Mercury at Superior Conjunction
09	16:25	Moon at Apogee: 405167 km
11	21:39	Spica 2.4°S of Moon
14	10:43	FULL MOON
21	18:29	LAST QUARTER MOON
22	05:01	Jupiter 4.8°S of Moon
25	03:25	Venus 2.0°S of Moon
25	13:01	Moon at Perigee: 365415 km
28	11:36	NEW MOON
29	17	Venus at Greatest Elong: 45.9°W
30	14:15	Saturn 4.4°N of Moon
Jun 04	05	Mercury 2.4°N of Saturn
04	23:39	FIRST QUARTER MOON
06	10:13	Moon at Apogee: 404291 km
07	16	Mercury at Greatest Elong: 23.8°E
08	05:09	Spica 2.5°S of Moon
12	23:19	FULL MOON
18	13:20	Jupiter 4.6°S of Moon
19	23:29	LAST QUARTER MOON
21	01:01	Summer Solstice
21	01:28	Moon at Perigee: 369518 km
23	02	Uranus in Conjunction with Sun
23	11:05	Venus 0.1°S of Moon: Occn.
25	00:24	Aldebaran 1.5°S of Moon
26	21:07	NEW MOON
27	23	Mars at Opposition

Date	GMT (h:m)	Event
Jul 02	08	Saturn in Conjunction with Sun
03	22	Earth at Aphelion: 1.01669 AU
04	04:55	Moon at Apogee: 404093 km
04	11	Mercury at Inferior Conjunction
04	17:12	FIRST QUARTER MOON
05	13:01	Spica 2.8°S of Moon
12	09:29	FULL MOON
12	22:47	Venus 3.1°N of Aldebaran
15	18:27	Jupiter 4.5°S of Moon
16	09:27	Moon at Perigee: 367673 km
19	04:07	LAST QUARTER MOON
22	06:52	Aldebaran 1.4°S of Moon
23	02:28	Venus 2.7°N of Moon
24	17:13	Mercury 2.6°N of Moon
24	19:55	Saturn 4.6°N of Moon
25	15	Mercury at Greatest Elong: 20.0°W
26	08:13	NEW MOON
26	17	Mercury 1.5°S of Saturn
28	05	Delta-Aquarid Meteor Shower
31	23:13	Moon at Apogee: 404734 km
Aug 01	20:40	Spica 3.0°S of Moon
03	10:26	FIRST QUARTER MOON
10	18:08	FULL MOON
11	22:26	Jupiter 4.7°S of Moon
12	21	Perseid Meteor Shower
12	21:22	Moon at Perigee: 362710 km
13	09	Venus 0.3°S of Saturn
17	09:43	LAST QUARTER MOON
18	12:13	Aldebaran 1.2°S of Moon
20	10	Mercury at Superior Conjunction
24	21:40	NEW MOON
25	04	Jupiter at Opposition
28	15:27	Moon at Apogee: 405762 km
29	03:40	Spica 3.2°S of Moon
Sep 02	02:24	FIRST QUARTER MOON
08	03:10	Jupiter 5.0°S of Moon
09	02:20	FULL MOON
10	01:49	Moon at Perigee: 358590 km
14	18:16	Aldebaran 1.0°S of Moon
15	17:34	LAST QUARTER MOON
18	20:29	Venus 0.4°N of Regulus
22	16:52	Autumnal Equinox
23	05:56	Mercury 0.4°N of Spica
23	13:40	NEW MOON
23	13:53	Partial Solar Eclipse; mag=0.689
25	01:33	Moon at Apogee: 406444 km
25	10:00	Spica 3.2°S of Moon
25	16:31	Mercury 3.4°S of Moon

Date	GMT (h:m)	Event
Oct 01	16:33	FIRST QUARTER MOON
03	09	Mars at Perihelion: 1.38124 AU
04	16	Mercury at Greatest Elong: 25.7°E
08	10:55	Total Lunar Eclipse; mag=1.350
08	10:58	FULL MOON
08	12:11	Moon at Perigee: 356825 km
11	16	Neptune at Opposition
12	02:32	Aldebaran 1.0°S of Moon
15	04:47	LAST QUARTER MOON
21	13	Orionid Meteor Shower
21	16:30	Venus 1.8°N of Moon
22	03:15	Moon at Apogee: 406439 km
23	07:28	NEW MOON
28	13	Mercury at Inferior Conjunction
31	04:46	FIRST QUARTER MOON
Nov 01	12:14	Venus 3.2°N of Spica
05	14	S Taurid Meteor Shower
05	23:57	Moon at Perigee: 358102 km
06	20:32	FULL MOON
08	12:59	Aldebaran 1.1°S of Moon
11	12:32	Saturn 4.8°N of Moon
12	13	N Taurid Meteor Shower
13	14	Mercury at Greatest Elong: 19.2°W
13	20:09	LAST QUARTER MOON
14	14:42	Regulus 4.1°N of Moon
17	19	Leonid Meteor Shower
18	10:41	Moon at Apogee: 405836 km
18	22:30	Spica 3.3°S of Moon
20	11:14	Mercury 0.6°S of Moon: Occn.
22	01:39	NEW MOON
29	03:59	Jupiter 4.7°S of Moon
29	15:15	FIRST QUARTER MOON
Dec 01	10	Mars 0.2°S of Jupiter
04	08:06	Moon at Perigee: 362272 km
05	23:58	Aldebaran 1.1°S of Moon
06	07:22	FULL MOON
08	20:26	Saturn 4.7°N of Moon
11	22:38	Regulus 3.9°N of Moon
13	15:28	LAST QUARTER MOON
14	09	Geminid Meteor Shower
16	03:29	Moon at Apogee: 404906 km
16	05:36	Spica 3.5°S of Moon
21	13:45	Winter Solstice
21	18:47	NEW MOON
22	17	Ursid Meteor Shower
23	15	Mercury at Superior Conjunction
25	15	Uranus at Opposition
26	16:14	Jupiter 4.0°S of Moon
27	17:15	Mars 2.4°S of Moon
29	00:20	FIRST QUARTER MOON

Almanac of Astronomical Events for 2034

Date	GMT (h:m)	Event
Jan 01	00:06	Moon at Perigee: 367922 km
02	09:17	Aldebaran 1.0°S of Moon
03	23	Quadrantid Meteor Shower
04	02	Venus at Superior Conjunction
04	04	Earth at Perihelion: 0.98329 AU
04	19:47	FULL MOON
05	02:56	Saturn 4.6°N of Moon
08	02	Saturn at Opposition
08	07:42	Regulus 3.6°N of Moon
12	13:17	LAST QUARTER MOON
12	13:24	Spica 3.8°S of Moon
13	00:22	Moon at Apogee: 404300 km
20	10:02	NEW MOON
21	21:40	Mercury 4.8°S of Moon
23	07:41	Jupiter 3.4°S of Moon
25	08:02	Mars 0.3°N of Moon: Occn.
25	21:19	Moon at Perigee: 369926 km
26	22	Mercury at Greatest Elong: 18.5°E
27	08:32	FIRST QUARTER MOON
29	15:58	Aldebaran 0.8°S of Moon
Feb 01	07:29	Saturn 4.8°N of Moon
03	10:05	FULL MOON
04	16:36	Regulus 3.5°N of Moon
08	21:27	Spica 4.0°S of Moon
09	21:37	Moon at Apogee: 404475 km
11	09	Mercury at Inferior Conjunction
11	11:09	LAST QUARTER MOON
18	23:10	NEW MOON
21	15:30	Moon at Perigee: 364926 km
22	23:10	Mars 2.7°N of Moon
25	16:34	FIRST QUARTER MOON
25	21:17	Aldebaran 0.7°S of Moon
Mar 04	00:10	Regulus 3.5°N of Moon
05	02:10	FULL MOON
09	12	Mercury at Greatest Elong: 27.4°W
09	16:05	Moon at Apogee: 405268 km
10	15	Jupiter in Conjunction with Sun
13	06:45	LAST QUARTER MOON
18	15:54	Mercury 4.7°S of Moon
20	10:14	NEW MOON
20	10:17	Total Solar Eclipse; mag=1.046
20	13:18	Vernal Equinox
21	17:33	Venus 0.7°N of Moon: Occn.
21	18:12	Moon at Perigee: 359963 km
23	15:31	Mars 4.6°N of Moon
25	03:43	Aldebaran 0.6°S of Moon
27	01:18	FIRST QUARTER MOON
31	06:14	Regulus 3.5°N of Moon

Date	GMT (h:m)	Event
Apr 03	03	Mercury 1.3°S of Jupiter
03	19:06	Pen. Lunar Eclipse; mag=0.855
03	19:19	FULL MOON
06	03:44	Moon at Apogee: 406054 km
08	10:24	Mars 3.2°S of Pleiades
09	13	Neptune in Conjunction with Sun
11	22:45	LAST QUARTER MOON
16	20:42	Jupiter 1.5°S of Moon
18	19:26	NEW MOON
19	04:07	Moon at Perigee: 357338 km
20	12:55	Venus 4.7°N of Moon
21	12:35	Aldebaran 0.7°S of Moon
21	20	Mercury at Superior Conjunction
22	20	Lyrid Meteor Shower
23	15:39	Venus 3.4°S of Pleiades
25	11:35	FIRST QUARTER MOON
27	11:51	Regulus 3.4°N of Moon
May 03	06:24	Moon at Apogee: 406350 km
03	12:16	FULL MOON
05	09	Eta-Aquarid Meteor Shower
05	11:29	Mercury 2.2°S of Pleiades
11	10:56	LAST QUARTER MOON
12	07	Venus 0.4°N of Mars
14	16:00	Jupiter 0.9°S of Moon: Occn.
17	14:28	Moon at Perigee: 357645 km
18	03:12	NEW MOON
20	08	Mercury at Greatest Elong: 22.2°E
21	14:32	Saturn 4.9°N of Moon
24	18:25	Regulus 3.2°N of Moon
24	23:57	FIRST QUARTER MOON
30	11:23	Moon at Apogee: 406002 km
Jun 02	03:54	FULL MOON
03	23	Venus 2.0°N of Saturn
09	19:44	LAST QUARTER MOON
11	07:42	Jupiter 0.3°S of Moon: Occn.
14	11	Mercury at Inferior Conjunction
14	21:42	Moon at Perigee: 360644 km
15	09:36	Aldebaran 0.7°S of Moon
16	10:26	NEW MOON
18	05:58	Saturn 4.7°N of Moon
21	02:31	Regulus 3.0°N of Moon
21	06:45	Summer Solstice
23	14:35	FIRST QUARTER MOON
26	21	Mars 1.1°N of Saturn
27	00:19	Moon at Apogee: 405126 km
28	00	Uranus in Conjunction with Sun

Date	GMT (h:m)	Event
Jul 01	17:44	FULL MOON
06	19	Earth at Aphelion: 1.01667 AU
07	21	Mercury at Greatest Elong: 21.3°W
08	18:56	Jupiter 0.1°N of Moon: Occn.
09	01:59	LAST QUARTER MOON
09	05:45	Venus 0.9°N of Regulus
12	18:24	Aldebaran 0.6°S of Moon
12	19:35	Moon at Perigee: 365365 km
14	07:05	Mercury 3.4°N of Moon
15	18:15	NEW MOON
17	10	Saturn in Conjunction with Sun
18	11:40	Regulus 2.8°N of Moon
19	07:49	Venus 2.3°N of Moon
23	07:05	FIRST QUARTER MOON
24	17:20	Moon at Apogee: 404334 km
28	11	Delta-Aquarid Meteor Shower
31	05:54	FULL MOON
Aug 04	08	Mercury at Superior Conjunction
05	01:58	Jupiter 0.3°N of Moon: Occn.
07	06:50	LAST QUARTER MOON
08	08:24	Moon at Perigee: 369503 km
09	00:59	Aldebaran 0.5°S of Moon
12	12:38	Saturn 4.4°N of Moon
12	19	Venus at Greatest Elong: 45.9°E
13	03	Perseid Meteor Shower
14	03:53	NEW MOON
17	21:23	Venus 3.5°S of Moon
19	06	Mars in Conjunction with Sun
21	12:02	Moon at Apogee: 404240 km
22	00:43	FIRST QUARTER MOON
29	16:49	FULL MOON
Sep 01	06:19	Jupiter 0.1°N of Moon: Occn.
02	01:15	Venus 1.4°S of Spica
02	15:31	Moon at Perigee: 367629 km
05	06:21	Aldebaran 0.3°S of Moon
05	11:41	LAST QUARTER MOON
09	01:04	Saturn 4.4°N of Moon
11	04:25	Regulus 2.7°N of Moon
11	20	Mars at Aphelion: 1.66606 AU
12	16:14	NEW MOON
12	16:18	Annular Solar Eclipse; mag=0.974
17	04	Mercury at Greatest Elong: 26.6°E
18	07:04	Moon at Apogee: 404970 km
20	08:25	Mercury 0.9°S of Spica
20	18:39	FIRST QUARTER MOON
22	22:41	Autumnal Equinox
28	02:46	Partial Lunar Eclipse; mag=0.014
28	02:57	FULL MOON
28	10:13	Jupiter 0.2°S of Moon: Occn.
30	04:10	Moon at Perigee: 362430 km

Date	GMT (h:m)	Event
Oct 02	00	Jupiter at Opposition
02	12:32	Aldebaran 0.4°S of Moon
04	18:05	LAST QUARTER MOON
06	11:00	Saturn 4.3°N of Moon
08	10:34	Regulus 2.7°N of Moon
10	18:17	Mars 0.5°N of Moon: Occn.
12	07:33	NEW MOON
12	11	Mercury at Inferior Conjunction
14	04	Neptune at Opposition
16	00:06	Moon at Apogee: 406022 km
20	12:03	FIRST QUARTER MOON
21	17	Venus at Inferior Conjunction
21	19	Orionid Meteor Shower
25	02	Mercury 4.9°S of Mars
25	15:31	Jupiter 0.5°S of Moon: Occn.
27	12:42	FULL MOON
28	01	Mercury at Greatest Elong: 18.5°W
28	10:16	Moon at Perigee: 358119 km
29	21:10	Aldebaran 0.5°S of Moon
Nov 01	21:39	Venus 1.6°S of Spica
02	19:13	Saturn 4.1°N of Moon
03	03:27	LAST QUARTER MOON
04	15:59	Regulus 2.6°N of Moon
05	20	S Taurid Meteor Shower
08	11:07	Mars 1.3°S of Moon
11	01:16	NEW MOON
12	09:37	Moon at Apogee: 406644 km
13	11	Venus 2.6°S of Mars
17	02:50	Mars 2.7°N of Spica
18	02	Leonid Meteor Shower
19	04:01	FIRST QUARTER MOON
21	22:45	Jupiter 0.4°S of Moon: Occn.
22	20:05	Venus 2.6°N of Spica
25	22:06	Moon at Perigee: 356448 km
25	22:32	FULL MOON
26	08:07	Aldebaran 0.5°S of Moon
30	03:03	Saturn 3.9°N of Moon
Dec 01	22:26	Regulus 2.3°N of Moon
02	17	Mercury at Superior Conjunction
02	16:46	LAST QUARTER MOON
05	17	Jupiter at Perihelion: 4.95268 AU
06	18:30	Venus 1.2°S of Moon
07	05:25	Mars 3.0°S of Moon
09	10:00	Moon at Apogee: 406607 km
10	20:14	NEW MOON
14	15	Geminid Meteor Shower
18	17:45	FIRST QUARTER MOON
19	07:40	Jupiter 0.1°N of Moon: Occn.
21	19:35	Winter Solstice
22	23	Ursid Meteor Shower
23	19:28	Aldebaran 0.5°S of Moon
24	10:34	Moon at Perigee: 358118 km
25	08:54	FULL MOON
27	11:07	Saturn 3.9°N of Moon
29	07:10	Regulus 2.1°N of Moon
30	06	Uranus at Opposition
30	10	Venus 2.7°N of Mars

Almanac of Astronomical Events for 2035

Date	GMT (h:m)	Event
Jan 01	09	Venus at Greatest Elong: 46.9°W
01	10:01	LAST QUARTER MOON
03	01	Earth at Perihelion: 0.98333 AU
04	05	Quadrantid Meteor Shower
05	01:59	Mars 4.3°S of Moon
05	07:54	Venus 1.5°S of Moon
05	18:52	Moon at Apogee: 405976 km
09	15:03	NEW MOON
10	08	Mercury at Greatest Elong: 19.1°E
11	09:28	Mercury 3.6°S of Moon
15	18:28	Jupiter 0.7°N of Moon: Occn.
17	04:45	FIRST QUARTER MOON
20	04:52	Aldebaran 0.4°S of Moon
21	18:05	Moon at Perigee: 362707 km
22	04	Saturn at Opposition
23	18:37	Saturn 4.0°N of Moon
23	20:16	FULL MOON
25	17:32	Regulus 1.9°N of Moon
26	05	Mercury at Inferior Conjunction
31	06:02	LAST QUARTER MOON
Feb 02	12:48	Moon at Apogee: 405000 km
04	11:21	Venus 2.8°S of Moon
06	14:11	Mercury 1.2°S of Moon
08	08:22	NEW MOON
12	07:57	Jupiter 1.4°N of Moon
15	13:17	FIRST QUARTER MOON
16	11:29	Aldebaran 0.2°S of Moon
18	05:31	Moon at Perigee: 368326 km
19	21	Mercury at Greatest Elong: 26.4°W
20	00:29	Saturn 4.2°N of Moon
22	03:34	Regulus 1.9°N of Moon
22	08:54	FULL MOON
22	09:05	Pen. Lunar Eclipse; mag=0.965
Mar 02	03:01	LAST QUARTER MOON
02	09:34	Moon at Apogee: 404374 km
06	19:01	Venus 2.9°S of Moon
08	06:26	Mercury 3.7°S of Moon
09	23:05	Annular Solar Eclipse; mag=0.992
09	23:09	NEW MOON
12	00:45	Jupiter 2.0°N of Moon
15	01:35	Moon at Perigee: 369413 km
15	16:50	Aldebaran 0.2°S of Moon
16	20:15	FIRST QUARTER MOON
19	05:08	Saturn 4.3°N of Moon
20	19:03	Vernal Equinox
21	11:33	Regulus 1.9°N of Moon
23	22:42	FULL MOON
30	05:37	Moon at Apogee: 404557 km
31	23:06	LAST QUARTER MOON

Date	GMT (h:m)	Event
Apr 05	19	Mercury at Superior Conjunction
05	23:16	Venus 1.3°S of Moon
08	10:58	NEW MOON
11	01:12	Moon at Perigee: 364449 km
11	23:23	Aldebaran 0.3°S of Moon
12	02	Neptune in Conjunction with Sun
15	02:55	FIRST QUARTER MOON
15	10:41	Saturn 4.2°N of Moon
17	03	Jupiter in Conjunction with Sun
17	17:25	Regulus 1.8°N of Moon
22	13:21	FULL MOON
23	02	Lyrid Meteor Shower
26	22:33	Moon at Apogee: 405356 km
30	00:21	Mercury 1.4°S of Pleiades
30	16:54	LAST QUARTER MOON
30	22:01	Mars 4.3°S of Moon
May 02	09	Mercury at Greatest Elong: 20.8°E
05	16	Eta-Aquarid Meteor Shower
05	22:31	Venus 1.5°N of Moon
07	20:04	NEW MOON
09	03:09	Moon at Perigee: 359785 km
09	08:17	Aldebaran 0.4°S of Moon
12	19:21	Saturn 3.9°N of Moon
14	10:28	FIRST QUARTER MOON
14	22:46	Regulus 1.6°N of Moon
17	21	Venus 0.5°S of Jupiter
22	04:26	FULL MOON
24	09:19	Moon at Apogee: 406138 km
25	05	Mercury at Inferior Conjunction
29	15:45	Mars 3.6°S of Moon
30	07:31	LAST QUARTER MOON
Jun 03	14:11	Jupiter 3.3°N of Moon
04	19:43	Venus 3.9°N of Moon
05	01:53	Mercury 1.1°N of Moon: Occn.
06	03:21	NEW MOON
06	11:36	Moon at Perigee: 357357 km
07	05	Mercury 3.0°S of Venus
09	07:48	Saturn 3.6°N of Moon
11	05:28	Regulus 1.3°N of Moon
12	19:50	FIRST QUARTER MOON
19	18	Mercury at Greatest Elong: 22.9°W
20	12:30	Moon at Apogee: 406401 km
20	19:37	FULL MOON
21	12:33	Summer Solstice
23	09:10	Mercury 2.4°N of Aldebaran
27	03:36	Mars 3.2°S of Moon
28	18:43	LAST QUARTER MOON

Date	GMT (h:m)	Event
Jul 01	08:44	Jupiter 3.7°N of Moon
02	23	Uranus in Conjunction with Sun
03	05:15	Aldebaran 0.4°S of Moon
04	05:50	Mercury 4.0°N of Moon
04	21:00	Moon at Perigee: 357717 km
05	09:59	NEW MOON
05	19	Earth at Aphelion: 1.01674 AU
06	23:02	Saturn 3.2°N of Moon
08	14:14	Regulus 1.1°N of Moon
12	07:33	FIRST QUARTER MOON
17	17:31	Moon at Apogee: 406005 km
19	13	Mercury at Superior Conjunction
20	10:37	FULL MOON
25	05:54	Mars 3.4°S of Moon
28	02:55	LAST QUARTER MOON
28	17	Delta-Aquarid Meteor Shower
28	23:31	Jupiter 3.9°N of Moon
30	14:06	Aldebaran 0.3°S of Moon
Aug 01	08	Saturn in Conjunction with Sun
02	04:06	Moon at Perigee: 360719 km
03	17:12	NEW MOON
04	23:17	Mercury 2.0°N of Moon
05	00:23	Regulus 1.0°N of Moon
05	13:33	Mercury 0.7°N of Regulus
09	18	Venus at Superior Conjunction
10	21:52	FIRST QUARTER MOON
13	09	Perseid Meteor Shower
14	06:09	Moon at Apogee: 405130 km
19	01:00	FULL MOON
19	01:11	Partial Lunar Eclipse; mag=0.104
21	12	Mars at Perihelion: 1.38141 AU
21	17:03	Mars 4.1°S of Moon
25	09:38	Jupiter 4.0°N of Moon
26	09:08	LAST QUARTER MOON
26	20:44	Aldebaran 0.2°S of Moon
30	02:28	Moon at Perigee: 365533 km
30	16	Mercury at Greatest Elong: 27.3°E
31	06:28	Saturn 2.7°N of Moon
Sep 02	01:55	Total Solar Eclipse; mag=1.032
02	01:59	NEW MOON
09	14:47	FIRST QUARTER MOON
10	23:26	Moon at Apogee: 404365 km
15	17	Mars at Opposition
17	13:22	Mars 4.2°S of Moon
17	14:23	FULL MOON
21	15:26	Jupiter 3.9°N of Moon
23	02:08	Aldebaran 0.2°S of Moon
23	04:39	Autumnal Equinox
24	14:39	LAST QUARTER MOON
25	13:37	Moon at Perigee: 369772 km
26	03	Mercury at Inferior Conjunction
27	19:09	Saturn 2.5°N of Moon
28	18:33	Regulus 1.0°N of Moon

Date	GMT (h:m)	Event
Oct 01	13:07	NEW MOON
08	19:01	Moon at Apogee: 404310 km
09	09:49	FIRST QUARTER MOON
11	17	Mercury at Greatest Elong: 18.0°W
14	12:35	Mars 2.5°S of Moon
16	17	Neptune at Opposition
17	02:35	FULL MOON
18	18:52	Jupiter 3.7°N of Moon
20	08:23	Aldebaran 0.4°S of Moon
20	19:38	Moon at Perigee: 367444 km
22	02	Orionid Meteor Shower
23	20:57	LAST QUARTER MOON
25	04:41	Saturn 2.2°N of Moon
26	00:37	Regulus 0.9°N of Moon
31	02:59	NEW MOON
Nov 05	15:01	Moon at Apogee: 405052 km
06	02	S Taurid Meteor Shower
08	05	Jupiter at Opposition
08	05:50	FIRST QUARTER MOON
08	18:10	Venus 3.8°N of Antares
11	04:23	Mars 0.1°N of Moon: Occn.
12	02	Mercury at Superior Conjunction
13	01	N Taurid Meteor Shower
14	22:33	Jupiter 3.6°N of Moon
15	13:49	FULL MOON
16	17:05	Aldebaran 0.5°S of Moon
17	11:30	Moon at Perigee: 361943 km
18	08	Leonid Meteor Shower
21	12:01	Saturn 1.9°N of Moon
22	05:16	LAST QUARTER MOON
22	05:51	Regulus 0.6°N of Moon
29	19:38	NEW MOON
Dec 03	08:09	Moon at Apogee: 406041 km
08	01:05	FIRST QUARTER MOON
09	08:48	Mars 2.7°N of Moon
12	04:12	Jupiter 3.7°N of Moon
14	04:00	Aldebaran 0.5°S of Moon
14	21	Geminid Meteor Shower
15	00:33	FULL MOON
15	19:38	Moon at Perigee: 357747 km
18	19:05	Saturn 1.8°N of Moon
19	12:38	Regulus 0.3°N of Moon
21	16:29	LAST QUARTER MOON
22	01:31	Winter Solstice
23	05	Ursid Meteor Shower
24	13	Mercury at Greatest Elong: 20.0°E
29	14:31	NEW MOON
30	15:44	Moon at Apogee: 406575 km
31	06:27	Mercury 2.7°S of Moon

Almanac of Astronomical Events for 2036

Date	GMT (h:m)	Event
Jan 01	23:41	Venus 2.8°S of Moon
03	22	Uranus at Opposition
04	11	Quadrantid Meteor Shower
05	15	Earth at Perihelion: 0.98332 AU
06	17:48	FIRST QUARTER MOON
08	12:06	Jupiter 4.1°N of Moon
10	07	Mercury at Inferior Conjunction
10	15:07	Aldebaran 0.4°S of Moon
13	08:47	Moon at Perigee: 356519 km
13	11:16	FULL MOON
15	03:03	Saturn 1.9°N of Moon
15	22:15	Regulus 0.2°N of Moon
20	06:46	LAST QUARTER MOON
26	06:01	Mercury 1.6°S of Moon
26	16:05	Moon at Apogee: 406498 km
28	10:17	NEW MOON
Feb 01	02:44	Venus 1.7°N of Moon
02	06	Mercury at Greatest Elong: 25.2°W
04	22:01	Jupiter 4.4°N of Moon
05	02	Saturn at Opposition
05	07:01	FIRST QUARTER MOON
07	00:12	Aldebaran 0.3°S of Moon
10	20:55	Moon at Perigee: 358657 km
11	11:12	Saturn 2.1°N of Moon
11	22:09	FULL MOON
11	22:12	Total Lunar Eclipse; mag=1.299
12	09:34	Regulus 0.1°N of Moon
18	02	Mars 1.9°N of Jupiter
18	23:47	LAST QUARTER MOON
23	03:15	Moon at Apogee: 405790 km
25	17:04	Mercury 2.6°S of Moon
27	04:45	Partial Solar Eclipse; mag=0.629
27	04:59	NEW MOON
Mar 03	10:18	Jupiter 4.5°N of Moon
05	06:41	Aldebaran 0.3°S of Moon
05	16:49	FIRST QUARTER MOON
09	18:10	Saturn 2.3°N of Moon
10	02:38	Moon at Perigee: 363380 km
10	20:08	Regulus 0.1°N of Moon
12	09:09	FULL MOON
14	23:05	Mars 2.8°S of Pleiades
19	08	Mercury at Superior Conjunction
19	18:39	LAST QUARTER MOON
20	01:02	Vernal Equinox
20	05	Venus at Greatest Elong: 46.2°E
21	21:41	Moon at Apogee: 404752 km
23	06	Venus 4.0°N of Jupiter
27	20:57	NEW MOON
31	01:27	Jupiter 4.5°N of Moon

Date	GMT (h:m)	Event
Apr 01	12:06	Aldebaran 0.5°S of Moon
03	22:38	Venus 0.0°S of Pleiades
04	00:03	FIRST QUARTER MOON
05	23:47	Saturn 2.2°N of Moon
06	09:36	Moon at Perigee: 368619 km
07	04:11	Regulus 0.0°N of Moon
10	20:22	FULL MOON
13	15	Neptune in Conjunction with Sun
13	21	Mercury at Greatest Elong: 19.6°E
18	14:06	LAST QUARTER MOON
18	17:40	Moon at Apogee: 404135 km
22	09	Lyrid Meteor Shower
26	09:33	NEW MOON
27	19:29	Jupiter 4.4°N of Moon
28	18:36	Aldebaran 0.7°S of Moon
May 01	08:26	Moon at Perigee: 369111 km
03	05:40	Saturn 2.0°N of Moon
03	05:54	FIRST QUARTER MOON
04	06	Mercury at Inferior Conjunction
04	09:54	Regulus 0.2°S of Moon
04	22	Eta-Aquarid Meteor Shower
10	08:09	FULL MOON
16	12:41	Moon at Apogee: 404366 km
18	08:39	LAST QUARTER MOON
23	23	Jupiter in Conjunction with Sun
24	01:01	Mercury 1.5°N of Moon
25	19:17	NEW MOON
28	09:17	Moon at Perigee: 364390 km
28	16:25	Mars 4.1°N of Moon
30	02	Venus at Inferior Conjunction
30	14:00	Saturn 1.6°N of Moon
31	08	Mercury at Greatest Elong: 24.6°W
31	15:13	Regulus 0.4°S of Moon
Jun 01	11:34	FIRST QUARTER MOON
08	21:02	FULL MOON
12	23	Mercury 0.2°S of Venus
13	05:06	Moon at Apogee: 405205 km
17	01:03	LAST QUARTER MOON
17	02	Mercury 0.8°S of Jupiter
20	18:31	Summer Solstice
21	21:51	Venus 1.8°N of Moon
22	12:50	Jupiter 4.2°N of Moon
22	13:12	Aldebaran 0.7°S of Moon
24	03:09	NEW MOON
25	10:31	Moon at Perigee: 359947 km
26	06:20	Mars 2.5°N of Moon
27	01:49	Saturn 1.2°N of Moon: Occn.
27	22:09	Regulus 0.6°S of Moon
30	18:13	FIRST QUARTER MOON

Date	GMT (h:m)	Event
Jul 02	22	Mercury at Superior Conjunction
03	22	Earth at Aphelion: 1.01667 AU
06	23	Uranus in Conjunction with Sun
08	11:19	FULL MOON
10	16:20	Moon at Apogee: 406005 km
14	02:43	Venus 1.1°N of Aldebaran
16	14:39	LAST QUARTER MOON
19	23:15	Aldebaran 0.6°S of Moon
20	03	Mars 0.1°N of Saturn
20	06:21	Venus 0.5°N of Moon: Occn.
20	09:15	Jupiter 4.0°N of Moon
22	11	Mercury 0.3°N of Saturn
23	10:17	NEW MOON
23	10:31	Partial Solar Eclipse; mag=0.199
23	15	Mercury 0.0°N of Mars
23	18:38	Moon at Perigee: 357534 km
24	03	Venus 3.5°S of Jupiter
24	16:38	Saturn 0.8°N of Moon: Occn.
24	20:26	Mars 0.7°N of Moon: Occn.
24	22:27	Mercury 0.5°N of Moon: Occn.
25	07:24	Regulus 0.7°S of Moon
28	00	Delta-Aquarid Meteor Shower
28	20:27	Mercury 0.1°N of Regulus
29	21	Mars at Aphelion: 1.66611 AU
30	02:56	FIRST QUARTER MOON
Aug 05	12:08	Mars 0.6°N of Regulus
06	19:59	Moon at Apogee: 406243 km
07	02:49	FULL MOON
07	02:51	Total Lunar Eclipse; mag=1.454
08	09	Venus at Greatest Elong: 45.8°W
12	02	Mercury at Greatest Elong: 27.4°E
12	15	Perseid Meteor Shower
15	00	Saturn in Conjunction with Sun
15	01:36	LAST QUARTER MOON
16	07:51	Aldebaran 0.6°S of Moon
17	03:04	Jupiter 3.8°N of Moon
18	12:03	Venus 0.4°N of Moon: Occn.
21	04:27	Moon at Perigee: 357819 km
21	17:24	Partial Solar Eclipse; mag=0.862
21	17:35	NEW MOON
28	14:43	FIRST QUARTER MOON
Sep 03	00:30	Moon at Apogee: 405835 km
05	18:45	FULL MOON
08	09	Mercury at Inferior Conjunction
12	14:26	Aldebaran 0.7°S of Moon
13	10:29	LAST QUARTER MOON
13	16:32	Jupiter 3.5°N of Moon
17	01:15	Venus 0.2°S of Moon: Occn.
18	00:30	Saturn 0.2°N of Moon: Occn.
18	04:32	Regulus 0.7°S of Moon
18	12:38	Moon at Perigee: 360820 km
18	20:53	Mercury 2.7°S of Moon
20	01:51	NEW MOON
22	10:23	Autumnal Equinox
23	16	Mars in Conjunction with Sun
24	09	Mercury at Greatest Elong: 17.9°W
27	06:12	FIRST QUARTER MOON
30	13:16	Moon at Apogee: 404995 km
30	16	Venus 0.8°S of Saturn

Date	GMT (h:m)	Event
Oct 01	20:00	Venus 0.1°S of Regulus
05	10:15	FULL MOON
09	19:54	Aldebaran 0.9°S of Moon
11	00:46	Jupiter 3.2°N of Moon
12	18:09	LAST QUARTER MOON
14	03:59	Saturn 0.7°N of Regulus
15	12:54	Regulus 0.9°S of Moon
15	13:31	Saturn 0.1°S of Moon: Occn.
16	12:04	Moon at Perigee: 365813 km
16	17:36	Venus 1.7°S of Moon
18	05	Neptune at Opposition
19	11:50	NEW MOON
21	08	Orionid Meteor Shower
22	10	Mercury at Superior Conjunction
27	01:14	FIRST QUARTER MOON
28	07:33	Moon at Apogee: 404295 km
Nov 04	00:44	FULL MOON
05	08	S Taurid Meteor Shower
06	02:10	Aldebaran 1.0°S of Moon
07	04:49	Jupiter 3.1°N of Moon
11	01:28	LAST QUARTER MOON
11	18:21	Moon at Perigee: 370165 km
11	18:51	Regulus 1.1°S of Moon
11	22:56	Saturn 0.5°S of Moon: Occn.
12	08	N Taurid Meteor Shower
15	13:11	Venus 3.5°N of Spica
15	13:34	Venus 3.2°S of Moon
16	11:00	Mars 4.8°S of Moon
17	14	Leonid Meteor Shower
18	00:14	NEW MOON
25	04:24	Moon at Apogee: 404313 km
25	22:28	FIRST QUARTER MOON
Dec 03	10:37	Aldebaran 1.1°S of Moon
03	14:08	FULL MOON
04	07:28	Jupiter 3.2°N of Moon
06	13	Mercury at Greatest Elong: 21.1°E
07	01:52	Moon at Perigee: 367178 km
07	05	Venus 1.2°N of Mars
09	00:06	Regulus 1.4°S of Moon
09	05:39	Saturn 0.7°S of Moon: Occn.
10	09:18	LAST QUARTER MOON
12	15	Jupiter at Opposition
14	03	Geminid Meteor Shower
15	15:36	Venus 3.7°S of Moon
17	15:34	NEW MOON
21	07:12	Winter Solstice
22	11	Ursid Meteor Shower
23	01:08	Moon at Apogee: 405081 km
24	12	Mercury at Inferior Conjunction
25	19:44	FIRST QUARTER MOON
30	20:56	Aldebaran 1.0°S of Moon
31	11:30	Jupiter 3.5°N of Moon

Almanac of Astronomical Events for 2037

Date	GMT (h:m)	Event
Jan 02	02:35	FULL MOON
03	04	Earth at Perihelion: 0.98329 AU
03	11	Mercury 2.6°N of Venus
03	17	Quadrantid Meteor Shower
03	21:29	Moon at Perigee: 361542 km
05	07:16	Regulus 1.5°S of Moon
05	11:57	Saturn 0.7°S of Moon: Occn.
07	15	Uranus at Opposition
08	18:29	LAST QUARTER MOON
13	02:16	Mars 4.3°S of Moon
14	06:55	Mercury 1.6°S of Moon
14	15	Mercury at Greatest Elong: 23.7°W
14	23:47	Venus 2.5°S of Moon
16	09:34	NEW MOON
16	09:48	Partial Solar Eclipse; mag=0.705
19	17:43	Moon at Apogee: 406033 km
24	14:55	FIRST QUARTER MOON
27	07:16	Aldebaran 1.0°S of Moon
27	18:07	Jupiter 3.8°N of Moon
31	14:00	Total Lunar Eclipse; mag=1.207
31	14:04	FULL MOON
Feb 01	06:58	Moon at Perigee: 357596 km
01	17:21	Regulus 1.6°S of Moon
01	19:24	Saturn 0.5°S of Moon: Occn.
07	05:43	LAST QUARTER MOON
11	00:49	Mars 3.2°S of Moon
13	04:35	Saturn 1.1°N of Regulus
15	04:54	NEW MOON
15	23:45	Moon at Apogee: 406539 km
17	19	Saturn at Opposition
23	06:41	FIRST QUARTER MOON
23	15:45	Aldebaran 1.0°S of Moon
24	03:08	Jupiter 3.8°N of Moon
Mar 01	03:37	Saturn 0.2°S of Moon: Occn.
01	04:52	Regulus 1.6°S of Moon
01	19:48	Moon at Perigee: 356710 km
02	00:28	FULL MOON
02	08	Mercury at Superior Conjunction
08	19:25	LAST QUARTER MOON
12	00:56	Mars 1.7°S of Moon
15	01:18	Moon at Apogee: 406442 km
16	23:36	NEW MOON
20	06:50	Vernal Equinox
21	09	Venus at Superior Conjunction
22	22:06	Aldebaran 1.2°S of Moon
23	14:19	Jupiter 3.5°N of Moon
24	18:39	FIRST QUARTER MOON
27	20	Mercury at Greatest Elong: 18.8°E
28	11:13	Saturn 0.2°S of Moon: Occn.
28	15:19	Regulus 1.6°S of Moon
30	06:31	Moon at Perigee: 359008 km
31	09:53	FULL MOON

Date	GMT (h:m)	Event
Apr 07	11:25	LAST QUARTER MOON
10	02:06	Mars 0.0°S of Moon: Occn.
11	13:04	Moon at Apogee: 405679 km
15	03	Mercury at Inferior Conjunction
15	16:08	NEW MOON
16	04	Neptune in Conjunction with Sun
19	03:37	Aldebaran 1.4°S of Moon
20	03:46	Jupiter 3.1°N of Moon
22	15	Lyrid Meteor Shower
23	03:11	FIRST QUARTER MOON
24	17:40	Saturn 0.4°S of Moon: Occn.
24	23:11	Regulus 1.8°S of Moon
27	09:43	Moon at Perigee: 363607 km
29	18:54	FULL MOON
May 05	04	Eta-Aquarid Meteor Shower
07	04:56	LAST QUARTER MOON
09	03:52	Mars 1.5°N of Moon
09	06:32	Moon at Apogee: 404662 km
12	23:06	Mercury 1.9°N of Moon
13	00	Mercury at Greatest Elong: 26.2°W
15	05:54	NEW MOON
17	19:41	Jupiter 2.6°N of Moon
22	00:06	Saturn 0.7°S of Moon: Occn.
22	04:50	Regulus 2.1°S of Moon
22	09:08	FIRST QUARTER MOON
24	14:45	Moon at Perigee: 368540 km
29	04:24	FULL MOON
Jun 03	08	Venus 1.1°N of Jupiter
05	22:49	LAST QUARTER MOON
06	01:22	Moon at Apogee: 404141 km
07	05:39	Mars 2.7°N of Moon
13	17:10	NEW MOON
15	11:35	Venus 2.5°N of Moon
17	09	Mercury at Superior Conjunction
18	08:21	Saturn 1.1°S of Moon: Occn.
18	10:13	Regulus 2.3°S of Moon
18	16:37	Moon at Perigee: 368923 km
20	13:45	FIRST QUARTER MOON
21	00:22	Summer Solstice
27	15:20	FULL MOON
29	14	Jupiter in Conjunction with Sun

Date	GMT (h:m)	Event
Jul 03	19:50	Moon at Apogee: 404478 km
05	06:37	Saturn 1.0°N of Regulus
05	16:00	LAST QUARTER MOON
06	06:08	Mars 3.2°N of Moon
06	12	Earth at Aphelion: 1.01666 AU
08	07	Mars at Perihelion: 1.38132 AU
10	03:11	Aldebaran 1.5°S of Moon
11	23	Uranus in Conjunction with Sun
13	02:32	NEW MOON
13	02:39	Total Solar Eclipse; mag=1.041
14	21:51	Mercury 1.0°S of Moon: Occn.
15	07:22	Venus 0.6°S of Moon: Occn.
15	16:48	Moon at Perigee: 364330 km
15	17:16	Regulus 2.3°S of Moon
15	19:38	Saturn 1.4°S of Moon
18	15	Mercury 5.0°N of Venus
19	18:31	FIRST QUARTER MOON
20	21:08	Venus 1.0°N of Regulus
22	10	Venus 0.0°N of Saturn
25	02:05	Mercury 1.3°S of Regulus
25	08	Mercury at Greatest Elong: 27.0°E
27	04:09	Partial Lunar Eclipse; mag=0.809
27	04:15	FULL MOON
27	04	Mercury 3.0°S of Saturn
28	06	Delta-Aquarid Meteor Shower
31	12:33	Moon at Apogee: 405394 km
Aug 04	02:55	Mars 3.0°N of Moon
04	07:51	LAST QUARTER MOON
06	12:29	Aldebaran 1.5°S of Moon
09	06:26	Jupiter 1.1°N of Moon: Occn.
11	10:41	NEW MOON
12	17:37	Moon at Perigee: 359796 km
12	22	Perseid Meteor Shower
13	23:07	Venus 4.0°S of Moon
18	01:00	FIRST QUARTER MOON
22	02	Mercury at Inferior Conjunction
25	19:09	FULL MOON
28	00:28	Moon at Apogee: 406214 km
29	07	Saturn in Conjunction with Sun
Sep 01	16:14	Mars 2.6°N of Moon
02	20:38	Aldebaran 1.6°S of Moon
02	22:03	LAST QUARTER MOON
04	18:35	Venus 1.4°N of Spica
06	01:39	Jupiter 0.5°N of Moon: Occn.
07	23	Mercury at Greatest Elong: 18.0°W
08	10:37	Mercury 2.4°S of Moon
08	13:19	Regulus 2.3°S of Moon
09	18:25	NEW MOON
09	22:15	Mercury 0.1°N of Regulus
10	02:13	Moon at Perigee: 357229 km
16	10:36	FIRST QUARTER MOON
22	16:13	Autumnal Equinox
24	03:44	Moon at Apogee: 406418 km
24	11:32	FULL MOON
29	16:33	Mars 2.4°N of Moon
30	03:09	Aldebaran 1.9°S of Moon

Date	GMT (h:m)	Event
Oct 02	10:29	LAST QUARTER MOON
03	17:31	Jupiter 0.0°N of Moon: Occn.
03	22	Mercury at Superior Conjunction
05	23:38	Regulus 2.5°S of Moon
06	17:50	Saturn 2.1°S of Moon
08	13:04	Moon at Perigee: 357513 km
09	02:34	NEW MOON
16	00:15	FIRST QUARTER MOON
16	04:56	Venus 1.3°N of Antares
20	17	Neptune at Opposition
21	07:33	Moon at Apogee: 406008 km
21	14	Orionid Meteor Shower
24	04:36	FULL MOON
25	02	Venus at Greatest Elong: 47.0°E
26	21:27	Mars 3.3°N of Moon
27	08:47	Aldebaran 2.1°S of Moon
31	04:22	Jupiter 0.4°S of Moon: Occn.
31	21:06	LAST QUARTER MOON
Nov 02	07:48	Regulus 2.7°S of Moon
03	07:14	Saturn 2.4°S of Moon
05	14	S Taurid Meteor Shower
05	22:15	Moon at Perigee: 360791 km
07	12:03	NEW MOON
11	02:56	Mercury 2.0°N of Antares
12	14	N Taurid Meteor Shower
13	22:08	Mars 3.7°S of Pleiades
14	17:59	FIRST QUARTER MOON
17	20	Leonid Meteor Shower
17	21:05	Moon at Apogee: 405184 km
19	07	Mercury at Greatest Elong: 22.4°E
19	08	Mars at Opposition
22	21:35	FULL MOON
23	15:03	Aldebaran 2.2°S of Moon
27	10:04	Jupiter 0.6°S of Moon: Occn.
29	13:39	Regulus 2.9°S of Moon
30	06:06	LAST QUARTER MOON
30	16:48	Saturn 2.6°S of Moon
Dec 03	21:37	Moon at Perigee: 366169 km
06	23:38	NEW MOON
08	20	Mercury at Inferior Conjunction
09	15:10	Venus 1.6°S of Moon
14	09	Geminid Meteor Shower
14	14:42	FIRST QUARTER MOON
15	16:34	Moon at Apogee: 404504 km
20	22:59	Aldebaran 2.1°S of Moon
21	13:08	Winter Solstice
22	13:38	FULL MOON
22	18	Ursid Meteor Shower
24	12:47	Jupiter 0.4°S of Moon: Occn.
26	19:03	Regulus 3.1°S of Moon
27	23:09	Saturn 2.7°S of Moon
28	02	Mercury at Greatest Elong: 22.3°W
29	14:05	LAST QUARTER MOON
29	18:48	Moon at Perigee: 370285 km

Almanac of Astronomical Events for 2038

Date	GMT (h:m)	Event
Jan 03	05	Earth at Perihelion: 0.98335 AU
03	20:43	Mercury 1.4°S of Moon
03	23	Quadrantid Meteor Shower
04	01	Venus at Inferior Conjunction
05	13:41	NEW MOON
05	13:46	Annular Solar Eclipse; mag=0.973
12	08	Uranus at Opposition
12	13:57	Moon at Apogee: 404530 km
13	12:34	FIRST QUARTER MOON
14	20	Jupiter at Opposition
17	08:16	Aldebaran 2.1°S of Moon
20	15:41	Jupiter 0.0°S of Moon: Occn.
21	03:48	Pen. Lunar Eclipse; mag=0.900
21	04:00	FULL MOON
23	02:25	Regulus 3.1°S of Moon
24	04:38	Saturn 2.5°S of Moon
24	09:52	Moon at Perigee: 366508 km
27	22:00	LAST QUARTER MOON
Feb 04	05:52	NEW MOON
04	17:11	Mars 2.0°S of Pleiades
09	10:00	Moon at Apogee: 405275 km
12	09:30	FIRST QUARTER MOON
12	14	Mercury at Superior Conjunction
13	17:29	Aldebaran 2.2°S of Moon
16	20:47	Jupiter 0.3°N of Moon: Occn.
19	12:21	Regulus 3.0°S of Moon
19	16:09	FULL MOON
20	11:09	Saturn 2.3°S of Moon
21	08:05	Moon at Perigee: 360960 km
26	06:56	LAST QUARTER MOON
Mar 01	20:46	Venus 4.4°N of Moon
03	08	Saturn at Opposition
05	23:15	NEW MOON
09	00:39	Moon at Apogee: 406184 km
11	03	Mercury at Greatest Elong: 18.3°E
13	01:23	Aldebaran 2.4°S of Moon
13	13:38	Mars 4.6°N of Moon
14	03:42	FIRST QUARTER MOON
15	12	Venus at Greatest Elong: 46.6°W
16	04:29	Jupiter 0.2°N of Moon: Occn.
18	23:18	Regulus 3.1°S of Moon
19	18:49	Saturn 2.1°S of Moon
20	12:40	Vernal Equinox
21	02:09	FULL MOON
21	17:16	Moon at Perigee: 357387 km
27	17:36	LAST QUARTER MOON
27	21	Mercury at Inferior Conjunction
31	09:43	Venus 4.0°N of Moon
Apr 04	16:43	NEW MOON
05	05:32	Moon at Apogee: 406665 km
09	07:48	Aldebaran 2.7°S of Moon
11	02:58	Mars 2.8°N of Moon
12	14:25	Jupiter 0.1°S of Moon: Occn.
12	18:02	FIRST QUARTER MOON
15	09:08	Regulus 3.3°S of Moon
16	02:34	Saturn 2.2°S of Moon
18	17	Neptune in Conjunction with Sun
19	04:30	Moon at Perigee: 356842 km
19	10:36	FULL MOON
22	21	Lyrid Meteor Shower
24	20	Mercury at Greatest Elong: 27.3°W
26	06:15	LAST QUARTER MOON
30	09:51	Venus 3.8°N of Moon
May 01	22:14	Mercury 2.3°N of Moon
02	08:35	Moon at Apogee: 406509 km
04	09:19	NEW MOON
05	10	Eta-Aquarid Meteor Shower
06	13:34	Aldebaran 2.8°S of Moon
09	15:54	Mars 1.0°N of Moon: Occn.
10	02:19	Jupiter 0.6°S of Moon: Occn.
12	04:18	FIRST QUARTER MOON
12	16:39	Regulus 3.5°S of Moon
13	09:44	Saturn 2.5°S of Moon
17	13:35	Moon at Perigee: 359249 km
18	18:23	FULL MOON
23	02	Mars 1.0°N of Jupiter
25	20:43	LAST QUARTER MOON
29	20:16	Moon at Apogee: 405684 km
30	16:36	Venus 3.0°N of Moon
Jun 01	21	Mercury at Superior Conjunction
03	00:24	NEW MOON
06	16:15	Jupiter 1.2°S of Moon: Occn.
07	04:00	Mars 0.8°S of Moon: Occn.
08	22:20	Regulus 3.7°S of Moon
09	16:54	Saturn 2.7°S of Moon
10	11:11	FIRST QUARTER MOON
14	15:26	Moon at Perigee: 363752 km
17	00	Mars at Aphelion: 1.66611 AU
17	02:30	FULL MOON
17	02:44	Pen. Lunar Eclipse; mag=0.442
21	06:09	Summer Solstice
24	12:39	LAST QUARTER MOON
26	12:55	Moon at Apogee: 404680 km
30	01:31	Venus 1.4°N of Moon
30	03:02	Aldebaran 2.8°S of Moon
30	11	Mercury 0.5°N of Jupiter

Date	GMT (h:m)	Event
Jul 02	13:32	Annular Solar Eclipse; mag=0.991
02	13:32	NEW MOON
04	08:23	Jupiter 1.7°S of Moon
04	15:39	Mercury 2.2°S of Moon
04	20	Earth at Aphelion: 1.01669 AU
05	15:47	Mars 2.4°S of Moon
06	03:50	Regulus 3.7°S of Moon
07	01:26	Saturn 2.9°S of Moon
07	07	Mercury at Greatest Elong: 26.1°E
09	16:00	FIRST QUARTER MOON
11	19:32	Moon at Perigee: 368535 km
16	11:35	Pen. Lunar Eclipse; mag=0.500
16	11:48	FULL MOON
17	00	Uranus in Conjunction with Sun
17	15:30	Mars 0.6°N of Regulus
24	05:40	LAST QUARTER MOON
24	07:19	Moon at Apogee: 404203 km
27	11:15	Aldebaran 2.9°S of Moon
28	12	Delta-Aquarid Meteor Shower
30	07:32	Venus 0.6°S of Moon: Occn.
Aug 01	00:40	NEW MOON
02	10:48	Regulus 3.7°S of Moon
03	04:07	Mars 3.7°S of Moon
03	04	Jupiter in Conjunction with Sun
03	12:27	Saturn 3.0°S of Moon
04	03	Mercury at Inferior Conjunction
05	21:36	Moon at Perigee: 368833 km
07	20:21	FIRST QUARTER MOON
12	04	Mars 0.9°S of Saturn
13	04	Perseid Meteor Shower
14	22:57	FULL MOON
17	16	Mercury 2.9°S of Venus
21	01:57	Moon at Apogee: 404583 km
22	07	Mercury at Greatest Elong: 18.5°W
22	23:12	LAST QUARTER MOON
23	19:43	Aldebaran 3.0°S of Moon
26	19	Mercury 0.2°S of Jupiter
28	22:04	Jupiter 2.6°S of Moon
29	03:22	Mercury 2.6°S of Moon
30	10:13	NEW MOON
31	17:54	Mars 4.6°S of Moon
Sep 01	22:39	Moon at Perigee: 364106 km
06	01:51	FIRST QUARTER MOON
12	05	Saturn in Conjunction with Sun
13	12:24	FULL MOON
16	09	Mercury at Superior Conjunction
17	19:15	Moon at Apogee: 405516 km
20	03:32	Aldebaran 3.3°S of Moon
21	16:27	LAST QUARTER MOON
22	22:02	Autumnal Equinox
25	17:35	Jupiter 3.1°S of Moon
26	05:56	Regulus 3.7°S of Moon
28	18:57	NEW MOON
30	00:20	Moon at Perigee: 359444 km

Date	GMT (h:m)	Event
Oct 05	09:52	FIRST QUARTER MOON
13	04:22	FULL MOON
15	07:13	Moon at Apogee: 406282 km
17	10:16	Aldebaran 3.5°S of Moon
18	01	Venus at Superior Conjunction
21	08:23	LAST QUARTER MOON
21	20	Orionid Meteor Shower
23	05	Neptune at Opposition
23	11:07	Jupiter 3.5°S of Moon
23	15:39	Regulus 3.9°S of Moon
25	09:54	Saturn 3.3°S of Moon
28	03:53	NEW MOON
28	10:19	Moon at Perigee: 356944 km
Nov 01	07	Mars in Conjunction with Sun
01	22	Mercury at Greatest Elong: 23.7°E
03	21:24	FIRST QUARTER MOON
05	21	S Taurid Meteor Shower
10	08:27	Mercury 2.1°N of Antares
11	08:59	Moon at Apogee: 406419 km
11	22:27	FULL MOON
12	20	N Taurid Meteor Shower
13	16:19	Aldebaran 3.6°S of Moon
14	12:23	Jupiter 0.3°N of Regulus
18	02	Leonid Meteor Shower
19	22:10	LAST QUARTER MOON
19	23:25	Regulus 4.2°S of Moon
20	00:30	Jupiter 3.8°S of Moon
21	23:40	Saturn 3.5°S of Moon
23	03	Mercury at Inferior Conjunction
25	22:45	Moon at Perigee: 357562 km
26	13:47	NEW MOON
Dec 03	12:46	FIRST QUARTER MOON
08	13:35	Moon at Apogee: 405980 km
10	19	Mercury at Greatest Elong: 20.9°W
10	22:34	Aldebaran 3.6°S of Moon
11	17:30	FULL MOON
11	17:44	Pen. Lunar Eclipse; mag=0.805
14	15	Geminid Meteor Shower
17	05:15	Regulus 4.3°S of Moon
17	08:30	Jupiter 3.9°S of Moon
19	09:29	LAST QUARTER MOON
19	09:32	Saturn 3.4°S of Moon
21	19:01	Winter Solstice
23	00	Ursid Meteor Shower
24	08:23	Moon at Perigee: 361283 km
24	19:19	Mercury 0.8°S of Moon: Occn.
24	20:09	Mars 1.7°S of Moon
25	18	Mercury 0.8°N of Mars
26	00:59	Total Solar Eclipse; mag=1.027
26	01:02	NEW MOON
27	07:27	Venus 0.6°N of Moon: Occn.

Almanac of Astronomical Events for 2039

Date	GMT (h:m)	Event
Jan 02	07:37	FIRST QUARTER MOON
04	06	Quadrantid Meteor Shower
05	05:10	Moon at Apogee: 405107 km
05	06	Earth at Perihelion: 0.98331 AU
07	05:45	Aldebaran 3.6°S of Moon
10	11:45	FULL MOON
13	10:54	Regulus 4.2°S of Moon
13	12:09	Jupiter 3.6°S of Moon
15	15:43	Saturn 3.2°S of Moon
17	02	Uranus at Opposition
17	18:42	LAST QUARTER MOON
18	15:12	Jupiter 0.6°N of Regulus
21	05:35	Moon at Perigee: 366856 km
22	16:50	Mars 0.0°N of Moon: Occn.
24	13:36	NEW MOON
24	22	Mercury at Superior Conjunction
26	09:07	Venus 3.1°N of Moon
Feb 01	04:45	FIRST QUARTER MOON
02	01:38	Moon at Apogee: 404385 km
03	13:51	Aldebaran 3.7°S of Moon
09	03:39	FULL MOON
09	14:14	Jupiter 3.3°S of Moon
09	18:11	Regulus 4.1°S of Moon
11	20:19	Saturn 3.0°S of Moon
15	07	Jupiter at Opposition
15	17:30	Moon at Perigee: 370230 km
16	02:36	LAST QUARTER MOON
20	13:09	Mars 1.7°N of Moon
22	14	Mercury at Greatest Elong: 18.1°E
23	03:18	NEW MOON
25	13:46	Venus 4.6°N of Moon
Mar 01	22:42	Moon at Apogee: 404389 km
02	22:12	Aldebaran 4.0°S of Moon
03	02:15	FIRST QUARTER MOON
08	17:34	Jupiter 3.1°S of Moon
09	03:23	Regulus 4.2°S of Moon
10	09	Mercury at Inferior Conjunction
10	16:35	FULL MOON
11	01:36	Saturn 2.8°S of Moon
13	18:38	Moon at Perigee: 366000 km
16	16	Saturn at Opposition
17	10:08	LAST QUARTER MOON
20	18:32	Vernal Equinox
21	09:41	Mars 3.1°N of Moon
24	17:59	NEW MOON
27	22:28	Venus 4.4°N of Moon
29	17:29	Moon at Apogee: 405124 km

Date	GMT (h:m)	Event
Apr 01	21:55	FIRST QUARTER MOON
04	23:32	Jupiter 3.2°S of Moon
06	23	Mercury at Greatest Elong: 27.8°W
07	08:15	Saturn 2.8°S of Moon
09	02:53	FULL MOON
10	07:33	Venus 2.4°S of Pleiades
10	17:31	Moon at Perigee: 360831 km
15	18:07	LAST QUARTER MOON
19	07:29	Mars 4.0°N of Moon
21	04:54	Mercury 2.5°N of Moon
21	06	Neptune in Conjunction with Sun
23	03	Lyrid Meteor Shower
23	09:35	NEW MOON
26	06:54	Moon at Apogee: 406022 km
27	06:43	Venus 2.7°N of Moon
30	04:51	Pollux 4.1°N of Moon
May 01	14:07	FIRST QUARTER MOON
02	08:13	Jupiter 3.5°S of Moon
04	15:46	Saturn 3.0°S of Moon
05	16	Eta-Aquarid Meteor Shower
08	11:20	FULL MOON
09	01:48	Moon at Perigee: 357636 km
15	03:17	LAST QUARTER MOON
17	07	Mercury at Superior Conjunction
18	07:24	Mars 4.0°N of Moon
23	01:38	NEW MOON
23	12:06	Moon at Apogee: 406489 km
26	10	Mars at Perihelion: 1.38111 AU
27	05:34	Venus 0.2°N of Moon: Occn.
27	11:04	Pollux 4.0°N of Moon
29	06:35	Venus 3.8°S of Pollux
29	19:13	Jupiter 3.8°S of Moon
30	18	Venus at Greatest Elong: 45.4°E
31	02:24	FIRST QUARTER MOON
31	23:34	Saturn 3.2°S of Moon
Jun 06	12:01	Moon at Perigee: 357205 km
06	18:48	FULL MOON
06	18:53	Partial Lunar Eclipse; mag=0.885
13	14:16	LAST QUARTER MOON
16	09:08	Mars 3.2°N of Moon
18	23	Mercury at Greatest Elong: 24.7°E
19	15:56	Moon at Apogee: 406283 km
21	11:58	Summer Solstice
21	17:11	Annular Solar Eclipse; mag=0.945
21	17:21	NEW MOON
23	16:49	Pollux 4.0°N of Moon
23	20:20	Mercury 3.1°S of Moon
25	09:40	Venus 3.4°S of Moon
26	08:11	Jupiter 4.0°S of Moon
28	07:41	Saturn 3.3°S of Moon
29	11:17	FIRST QUARTER MOON

Date	GMT (h:m)	Event
Jul 04	20:31	Moon at Perigee: 359517 km
05	14	Earth at Aphelion: 1.01666 AU
06	02:03	FULL MOON
06	21:59	Jupiter 0.5°N of Regulus
13	03:38	LAST QUARTER MOON
15	11:08	Mars 1.8°N of Moon
16	08	Mercury at Inferior Conjunction
17	03:14	Moon at Apogee: 405452 km
21	07:54	NEW MOON
22	02	Uranus in Conjunction with Sun
23	23:09	Jupiter 4.2°S of Moon
25	16:51	Saturn 3.3°S of Moon
28	17:50	FIRST QUARTER MOON
28	18	Delta-Aquarid Meteor Shower
Aug 01	22:38	Moon at Perigee: 363901 km
04	09:56	FULL MOON
05	07	Mercury at Greatest Elong: 19.3°W
08	19	Venus at Inferior Conjunction
11	19:36	LAST QUARTER MOON
13	10	Perseid Meteor Shower
13	11:08	Mars 0.3°N of Moon: Occn.
13	19:35	Moon at Apogee: 404503 km
17	06:19	Pollux 3.9°N of Moon
19	20:50	NEW MOON
22	03:55	Saturn 3.2°S of Moon
26	23:16	FIRST QUARTER MOON
29	03:16	Moon at Perigee: 368737 km
30	13	Mercury at Superior Conjunction
Sep 02	19:23	FULL MOON
04	17	Jupiter in Conjunction with Sun
10	13:45	LAST QUARTER MOON
10	14:24	Moon at Apogee: 404110 km
11	06:51	Mars 1.0°S of Moon: Occn.
13	14:30	Pollux 3.7°N of Moon
18	08:23	NEW MOON
23	02:18	Moon at Perigee: 368956 km
23	03:50	Autumnal Equinox
25	04:53	FIRST QUARTER MOON
25	16	Saturn in Conjunction with Sun
27	06:05	Mercury 1.0°N of Spica

Date	GMT (h:m)	Event
Oct 02	07:23	FULL MOON
08	10:06	Moon at Apogee: 404575 km
09	12:28	Venus 2.0°S of Regulus
09	19:49	Mars 1.8°S of Moon
10	08:59	LAST QUARTER MOON
10	22:51	Pollux 3.5°N of Moon
15	06:40	Jupiter 4.4°S of Moon
15	10	Mercury at Greatest Elong: 25.0°E
16	08:44	Saturn 2.9°S of Moon
17	19:09	NEW MOON
19	04	Venus at Greatest Elong: 46.4°W
20	05:11	Moon at Perigee: 363947 km
22	02	Orionid Meteor Shower
24	11:51	FIRST QUARTER MOON
25	17	Neptune at Opposition
31	22:36	FULL MOON
Nov 02	12	Venus 0.2°S of Jupiter
05	04:29	Moon at Apogee: 405534 km
06	03	S Taurid Meteor Shower
06	22:03	Mars 1.7°S of Moon
07	06:32	Pollux 3.3°N of Moon
07	08	Mercury at Inferior Conjunction
09	03:46	LAST QUARTER MOON
12	01:43	Jupiter 4.4°S of Moon
12	19:23	Venus 3.7°S of Moon
13	00:32	Saturn 2.9°S of Moon
13	02	N Taurid Meteor Shower
15	03:18	Mercury 1.2°S of Moon: Occn.
15	09	Venus 0.6°S of Saturn
16	05:46	NEW MOON
17	09:19	Moon at Perigee: 359171 km
18	08	Leonid Meteor Shower
22	21:17	FIRST QUARTER MOON
23	21	Mercury at Greatest Elong: 19.8°W
30	16:49	FULL MOON
30	16:55	Partial Lunar Eclipse; mag=0.943
Dec 02	16:15	Moon at Apogee: 406245 km
04	07:13	Mars 0.6°S of Moon: Occn.
04	13:11	Pollux 3.3°N of Moon
08	20:44	LAST QUARTER MOON
09	17:43	Jupiter 4.3°S of Moon
10	14:30	Saturn 2.7°S of Moon
12	19:10	Venus 0.7°S of Moon: Occn.
14	21	Geminid Meteor Shower
15	16:22	Total Solar Eclipse; mag=1.036
15	16:32	NEW MOON
15	20:51	Moon at Perigee: 356818 km
22	00:41	Winter Solstice
22	10:01	FIRST QUARTER MOON
23	06	Ursid Meteor Shower
29	16:34	Moon at Apogee: 406355 km
30	12:37	FULL MOON
30	21:25	Mars 1.1°N of Moon: Occn.
31	19:17	Pollux 3.3°N of Moon

Almanac of Astronomical Events for 2040

Date	GMT (h:m)	Event
Jan 02	15	Mars at Opposition
03	12	Earth at Perihelion: 0.98329 AU
04	12	Quadrantid Meteor Shower
05	03	Mercury at Superior Conjunction
06	04:34	Jupiter 4.0°S of Moon
07	00:38	Saturn 2.4°S of Moon
07	11:05	LAST QUARTER MOON
11	17:32	Venus 1.9°N of Moon
13	10:03	Moon at Perigee: 357769 km
14	03:25	NEW MOON
21	02:21	FIRST QUARTER MOON
21	21	Uranus at Opposition
25	22:42	Moon at Apogee: 405896 km
26	10:20	Mars 1.9°N of Moon
28	01:39	Pollux 3.3°N of Moon
29	07:54	FULL MOON
Feb 02	09:51	Jupiter 3.7°S of Moon
03	06:39	Saturn 2.0°S of Moon
05	22:32	LAST QUARTER MOON
06	02	Mercury at Greatest Elong: 18.3°E
10	13:20	Venus 3.6°N of Moon
10	18:42	Moon at Perigee: 361749 km
12	14:24	NEW MOON
19	21:33	FIRST QUARTER MOON
21	13	Mercury at Inferior Conjunction
22	15:18	Moon at Apogee: 404988 km
22	16:55	Mars 1.1°N of Moon: Occn.
24	08:47	Pollux 3.2°N of Moon
28	00:59	FULL MOON
29	11:37	Jupiter 3.5°S of Moon
Mar 01	10:24	Saturn 1.7°S of Moon
06	03	Mercury 2.5°N of Venus
06	07:19	LAST QUARTER MOON
09	12:23	Moon at Perigee: 367227 km
11	09:26	Venus 4.1°N of Moon
13	01:46	NEW MOON
16	20	Jupiter at Opposition
19	07	Mercury at Greatest Elong: 27.7°W
20	00:11	Vernal Equinox
20	17:59	FIRST QUARTER MOON
21	11:22	Moon at Apogee: 404287 km
21	15:47	Mars 0.3°S of Moon: Occn.
22	16:39	Pollux 3.0°N of Moon
27	13:08	Jupiter 3.6°S of Moon
28	14:20	Saturn 1.7°S of Moon
28	15:11	FULL MOON
28	18	Saturn at Opposition

Date	GMT (h:m)	Event
Apr 02	00:25	Antares 4.3°S of Moon
03	20:39	Moon at Perigee: 369913 km
04	14:06	LAST QUARTER MOON
09	22:21	Mercury 2.4°N of Moon
11	14:00	NEW MOON
18	07:16	Moon at Apogee: 404364 km
18	23:26	Mars 1.8°S of Moon
19	00:43	Pollux 2.8°N of Moon
19	13:37	FIRST QUARTER MOON
22	09	Lyrid Meteor Shower
22	19	Neptune in Conjunction with Sun
23	17:00	Jupiter 3.8°S of Moon
24	19:49	Saturn 1.9°S of Moon
27	02:38	FULL MOON
29	07:40	Antares 4.1°S of Moon
30	04:22	Moon at Perigee: 365739 km
30	15	Mercury at Superior Conjunction
May 03	20:00	LAST QUARTER MOON
03	21	Mars at Aphelion: 1.66611 AU
04	22	Eta-Aquarid Meteor Shower
11	03:28	NEW MOON
11	03:42	Partial Solar Eclipse; mag=0.531
16	01:01	Moon at Apogee: 405171 km
16	08:20	Pollux 2.7°N of Moon
17	11:07	Mars 3.2°S of Moon
19	07:00	FIRST QUARTER MOON
21	00:17	Jupiter 4.0°S of Moon
22	02:56	Saturn 2.1°S of Moon
26	11:45	Total Lunar Eclipse; mag=1.535
26	11:47	FULL MOON
26	17:08	Antares 4.1°S of Moon
28	02:22	Moon at Perigee: 360810 km
30	13	Mercury at Greatest Elong: 23.2°E
31	02	Venus at Superior Conjunction
Jun 02	02:17	LAST QUARTER MOON
09	18:03	NEW MOON
11	09:42	Mercury 3.6°S of Moon
12	14:18	Moon at Apogee: 406115 km
12	15:09	Pollux 2.7°N of Moon
15	00:44	Mars 4.1°S of Moon
17	10:41	Jupiter 4.0°S of Moon
17	21:32	FIRST QUARTER MOON
18	11:15	Saturn 2.2°S of Moon
20	17:46	Summer Solstice
23	03:44	Antares 4.1°S of Moon
24	19:19	FULL MOON
25	09:33	Moon at Perigee: 357652 km
25	17	Mercury at Inferior Conjunction
25	22:20	Mars 0.6°N of Regulus

Date	GMT (h:m)	Event
Jul 01	10:18	LAST QUARTER MOON
05	19	Earth at Aphelion: 1.01673 AU
09	09:15	NEW MOON
09	20:13	Moon at Apogee: 406581 km
13	15:39	Mars 4.5°S of Moon
14	23:32	Jupiter 3.8°S of Moon
15	20:24	Saturn 2.1°S of Moon
17	09:16	FIRST QUARTER MOON
17	20	Mercury at Greatest Elong: 20.5°W
20	13:47	Antares 4.0°S of Moon
23	19:15	Moon at Perigee: 357113 km
24	02:05	FULL MOON
26	05	Uranus in Conjunction with Sun
28	00	Delta-Aquarid Meteor Shower
30	21:06	LAST QUARTER MOON
Aug 03	21:01	Venus 1.0°N of Regulus
05	23:43	Moon at Apogee: 406353 km
06	03:16	Pollux 2.7°N of Moon
08	00:26	NEW MOON
09	15:04	Venus 3.8°S of Moon
11	07:53	Mars 4.2°S of Moon
11	14:19	Jupiter 3.5°S of Moon
12	06:26	Saturn 1.8°S of Moon
12	16	Perseid Meteor Shower
13	06	Mercury at Superior Conjunction
15	18:36	FIRST QUARTER MOON
16	21:57	Antares 3.8°S of Moon
18	02	Mars 0.6°S of Jupiter
21	03:59	Moon at Perigee: 359359 km
22	09:09	FULL MOON
28	20:11	Pleiades 4.1°N of Moon
29	11:16	LAST QUARTER MOON
31	04	Mars 1.8°S of Saturn
Sep 01	17	Venus 0.2°S of Jupiter
02	09:43	Pollux 2.6°N of Moon
02	10:15	Moon at Apogee: 405547 km
06	15:13	NEW MOON
06	19	Venus 1.5°S of Saturn
07	21	Mercury 1.5°S of Jupiter
08	06:57	Jupiter 3.1°S of Moon
08	17:51	Saturn 1.5°S of Moon
08	20:58	Venus 3.1°S of Moon
09	01:36	Mars 3.3°S of Moon
11	13	Mercury 3.1°S of Saturn
13	04	Venus 0.2°N of Mars
13	04:04	Antares 3.5°S of Moon
14	02:07	FIRST QUARTER MOON
17	02:23	Venus 2.2°N of Spica
18	06:43	Moon at Perigee: 363882 km
20	08:32	Mars 2.1°N of Spica
20	16	Mercury 2.3°S of Mars
20	17:43	FULL MOON
20	20:54	Mercury 0.0°S of Spica
22	09:44	Autumnal Equinox
25	04:55	Pleiades 3.9°N of Moon
26	22	Mercury at Greatest Elong: 26.1°E
28	04:41	LAST QUARTER MOON
29	17:01	Pollux 2.4°N of Moon
30	02:50	Moon at Apogee: 404658 km

Date	GMT (h:m)	Event
Oct 04	19	Jupiter in Conjunction with Sun
06	05:26	NEW MOON
07	17	Saturn in Conjunction with Sun
07	20:46	Mars 1.9°S of Moon
08	23:52	Venus 1.3°S of Moon
10	09:31	Antares 3.3°S of Moon
10	20	Mercury 3.4°S of Mars
13	08:41	FIRST QUARTER MOON
15	11:20	Moon at Perigee: 368986 km
20	04:50	FULL MOON
21	08	Orionid Meteor Shower
21	10	Mercury at Inferior Conjunction
22	14:14	Pleiades 3.8°N of Moon
25	01:29	Venus 2.9°N of Antares
27	01:02	Pollux 2.2°N of Moon
27	05	Neptune at Opposition
27	22:34	Moon at Apogee: 404342 km
28	00:27	LAST QUARTER MOON
31	04	Mercury 4.0°S of Saturn
31	09	Mercury 4.1°S of Jupiter
Nov 01	07	Jupiter 1.1°S of Saturn
02	21:22	Jupiter 2.2°S of Moon
02	21:59	Saturn 1.0°S of Moon: Occn.
03	07:08	Mercury 0.8°S of Moon: Occn.
04	18:56	NEW MOON
04	19:08	Partial Solar Eclipse; mag=0.807
05	09	S Taurid Meteor Shower
06	05	Mercury at Greatest Elong: 18.9°W
06	16:22	Antares 3.3°S of Moon
07	21:29	Venus 0.7°N of Moon: Occn.
09	06:17	Moon at Perigee: 368784 km
11	15:23	FIRST QUARTER MOON
17	14	Leonid Meteor Shower
18	10	Jupiter at Aphelion: 5.45292 AU
18	19:03	Total Lunar Eclipse; mag=1.397
18	19:06	FULL MOON
18	22:51	Pleiades 3.8°N of Moon
23	09:11	Pollux 2.2°N of Moon
24	19:12	Moon at Apogee: 404851 km
26	21:07	LAST QUARTER MOON
30	07:08	Jupiter 3.0°N of Spica
30	13:22	Saturn 0.7°S of Moon: Occn.
30	17:30	Jupiter 1.7°S of Moon
Dec 04	07:33	NEW MOON
06	13:20	Moon at Perigee: 363352 km
07	13:45	Venus 2.4°N of Moon
10	23:30	FIRST QUARTER MOON
14	04	Geminid Meteor Shower
14	10	Mercury at Superior Conjunction
16	05:53	Pleiades 3.8°N of Moon
17	13	Mars in Conjunction with Sun
18	12:16	FULL MOON
20	16:44	Pollux 2.3°N of Moon
21	06:33	Winter Solstice
22	12	Ursid Meteor Shower
22	13:29	Moon at Apogee: 405778 km
26	17:02	LAST QUARTER MOON
28	03:06	Saturn 0.2°S of Moon: Occn.
28	11:38	Jupiter 1.1°S of Moon: Occn.
31	12:37	Antares 3.2°S of Moon

46

Almanac of Astronomical Events for 2041

Date	GMT (h:m)	Event
Jan 02	19:08	NEW MOON
03	18	Quadrantid Meteor Shower
03	19:39	Moon at Perigee: 358590 km
03	21	Earth at Perihelion: 0.98335 AU
05	09	Venus at Greatest Elong: 47.2°E
06	00:53	Venus 4.4°N of Moon
09	10:06	FIRST QUARTER MOON
12	11:38	Pleiades 3.6°N of Moon
16	23:20	Pollux 2.3°N of Moon
17	07:11	FULL MOON
18	23:16	Moon at Apogee: 406419 km
19	14	Mercury at Greatest Elong: 18.7°E
24	13:10	Saturn 0.3°N of Moon: Occn.
25	01:15	Jupiter 0.5°S of Moon: Occn.
25	10:33	LAST QUARTER MOON
25	17	Uranus at Opposition
27	22:55	Antares 3.1°S of Moon
Feb 01	05:43	NEW MOON
01	07:46	Moon at Perigee: 356610 km
04	04	Mercury at Inferior Conjunction
07	23:40	FIRST QUARTER MOON
08	17:42	Pleiades 3.4°N of Moon
13	05:20	Pollux 2.2°N of Moon
14	23:03	Moon at Apogee: 406498 km
16	02:21	FULL MOON
20	18:59	Saturn 0.7°N of Moon: Occn.
21	08:53	Jupiter 0.0°S of Moon: Occn.
24	00:29	LAST QUARTER MOON
24	06:54	Antares 2.8°S of Moon
Mar 01	06:52	Mars 4.2°N of Moon
01	17	Mercury at Greatest Elong: 27.0°W
01	19:56	Moon at Perigee: 358003 km
02	15:39	NEW MOON
08	01:31	Pleiades 3.1°N of Moon
09	15:51	FIRST QUARTER MOON
12	11:36	Pollux 2.0°N of Moon
14	07:06	Moon at Apogee: 405968 km
17	20:19	FULL MOON
18	07	Venus at Inferior Conjunction
19	02	Mercury 1.0°S of Mars
19	22:05	Saturn 0.8°N of Moon: Occn.
20	06:07	Vernal Equinox
20	11:26	Jupiter 0.1°N of Moon: Occn.
23	12:47	Antares 2.5°S of Moon
25	10:32	LAST QUARTER MOON
30	02:37	Moon at Perigee: 362174 km
30	04:54	Mars 3.7°N of Moon
Apr 01	01:29	NEW MOON
04	11:01	Pleiades 2.9°N of Moon
08	09:38	FIRST QUARTER MOON
08	18:52	Pollux 1.9°N of Moon
10	15	Saturn at Opposition
10	23:41	Moon at Apogee: 404998 km
12	14	Mars at Perihelion: 1.38129 AU
Apr 14	17	Mercury at Superior Conjunction
16	00:54	Saturn 0.7°N of Moon: Occn.
16	10	Jupiter at Opposition
16	11:52	Jupiter 0.1°S of Moon: Occn.
16	12:00	FULL MOON
19	18:25	Antares 2.4°S of Moon
22	15	Lyrid Meteor Shower
23	17:24	LAST QUARTER MOON
26	16:51	Moon at Perigee: 367410 km
28	03:15	Mars 2.4°N of Moon
30	11:46	NEW MOON
30	11:51	Total Solar Eclipse; mag=1.019
May 01	20:46	Pleiades 2.9°N of Moon
01	20:47	Mercury 0.8°N of Moon: Occn.
02	04:56	Mercury 1.9°S of Pleiades
05	05	Eta-Aquarid Meteor Shower
06	03:06	Pollux 1.8°N of Moon
08	03:54	FIRST QUARTER MOON
08	18:43	Moon at Apogee: 404308 km
12	08	Mercury at Greatest Elong: 21.6°E
12	17:11	Jupiter 3.3°N of Spica
13	05:18	Saturn 0.4°N of Moon: Occn.
13	13:39	Jupiter 0.5°S of Moon: Occn.
16	00:42	Partial Lunar Eclipse; mag=0.065
16	00:52	FULL MOON
17	01:27	Antares 2.4°S of Moon
22	01:18	Moon at Perigee: 369566 km
22	22:26	LAST QUARTER MOON
26	06:51	Venus 0.5°N of Moon: Occn.
27	01:24	Mars 0.5°N of Moon: Occn.
27	08	Venus at Greatest Elong: 45.9°W
29	22:56	NEW MOON
Jun 02	11:29	Pollux 1.9°N of Moon
05	13	Mercury at Inferior Conjunction
05	13:34	Moon at Apogee: 404429 km
06	21:40	FIRST QUARTER MOON
09	11:58	Saturn 0.3°N of Moon: Occn.
09	19:02	Jupiter 0.7°S of Moon: Occn.
13	10:18	Antares 2.5°S of Moon
14	10:59	FULL MOON
15	21	Jupiter 4.2°S of Saturn
17	11:48	Moon at Perigee: 365482 km
17	16:40	Mercury 1.2°N of Aldebaran
20	23:37	Summer Solstice
21	03:12	LAST QUARTER MOON
24	20:50	Venus 3.3°S of Moon
24	22:29	Mars 1.3°S of Moon
25	12:04	Pleiades 2.8°N of Moon
28	11:17	NEW MOON
28	15	Venus 2.0°S of Mars
29	19:09	Pollux 2.0°N of Moon
29	22	Mercury at Greatest Elong: 22.0°W

Date	GMT (h:m)	Event
Jul 03	06:46	Moon at Apogee: 405266 km
04	02	Earth at Aphelion: 1.01669 AU
06	14:12	FIRST QUARTER MOON
06	20:40	Saturn 0.4°N of Moon: Occn.
07	04:25	Jupiter 0.6°S of Moon: Occn.
10	20:10	Antares 2.4°S of Moon
12	12:55	Venus 3.2°N of Aldebaran
13	19:01	FULL MOON
15	09:14	Moon at Perigee: 360651 km
20	09:13	LAST QUARTER MOON
22	17:40	Pleiades 2.7°N of Moon
23	17:49	Mars 2.8°S of Moon
27	02:45	Jupiter 3.0°N of Spica
28	01:02	NEW MOON
28	06	Delta-Aquarid Meteor Shower
28	07	Mercury at Superior Conjunction
30	20:07	Moon at Apogee: 406212 km
31	08	Uranus in Conjunction with Sun
Aug 03	06:45	Saturn 0.7°N of Moon: Occn.
03	12:19	Spica 3.9°S of Moon
03	16:59	Jupiter 0.2°S of Moon: Occn.
05	04:53	FIRST QUARTER MOON
07	05:36	Antares 2.1°S of Moon
12	02:04	FULL MOON
12	16:02	Moon at Perigee: 357495 km
12	22	Perseid Meteor Shower
18	17:43	LAST QUARTER MOON
18	23:34	Pleiades 2.4°N of Moon
21	11:15	Mars 3.8°S of Moon
23	07:27	Pollux 1.9°N of Moon
26	16:16	NEW MOON
27	02:03	Moon at Apogee: 406632 km
28	18:25	Mercury 4.4°S of Moon
30	17:45	Saturn 1.0°N of Moon: Occn.
30	18:38	Spica 3.7°S of Moon
31	07:45	Jupiter 0.3°N of Moon: Occn.
Sep 03	13:25	Antares 1.9°S of Moon
03	17:19	FIRST QUARTER MOON
09	10	Mercury at Greatest Elong: 27.0°E
10	02:12	Moon at Perigee: 357004 km
10	09:24	FULL MOON
15	07:10	Pleiades 2.2°N of Moon
17	05:33	LAST QUARTER MOON
18	07:53	Venus 0.5°N of Regulus
19	02:59	Mars 4.3°S of Moon
19	13:20	Pollux 1.7°N of Moon
22	09:50	Mercury 0.7°S of Spica
22	15:27	Autumnal Equinox
22	22:53	Venus 3.8°S of Moon
23	04:59	Moon at Apogee: 406361 km
25	08:41	NEW MOON
27	00:28	Spica 3.6°S of Moon
27	05:40	Saturn 1.3°N of Moon
28	00:09	Jupiter 0.9°N of Moon: Occn.
30	19:29	Antares 1.7°S of Moon

Date	GMT (h:m)	Event
Oct 03	03:33	FIRST QUARTER MOON
05	06	Mercury at Inferior Conjunction
08	12:08	Moon at Perigee: 359445 km
09	18:03	FULL MOON
12	16:45	Pleiades 2.1°N of Moon
16	20:22	Pollux 1.6°N of Moon
16	21:05	LAST QUARTER MOON
17	16:41	Mars 4.2°S of Moon
19	14	Mercury 0.2°N of Venus
20	09	Saturn in Conjunction with Sun
20	16:08	Moon at Apogee: 405544 km
20	19	Mercury at Greatest Elong: 18.3°W
21	15	Orionid Meteor Shower
23	11:15	Mercury 0.1°S of Moon: Occn.
23	12:05	Venus 0.6°S of Moon: Occn.
25	01:30	NEW MOON
25	01:35	Annular Solar Eclipse; mag=0.947
27	20	Mercury 0.6°N of Venus
28	01:01	Antares 1.7°S of Moon
29	17	Neptune at Opposition
31	23:18	Venus 3.2°N of Spica
Nov 01	12:05	FIRST QUARTER MOON
04	04	Jupiter in Conjunction with Sun
05	15	S Taurid Meteor Shower
05	15:42	Moon at Perigee: 364299 km
08	04:34	Partial Lunar Eclipse; mag=0.170
08	04:43	FULL MOON
09	03:14	Pleiades 2.0°N of Moon
13	04:52	Pollux 1.6°N of Moon
15	02:35	Mars 3.4°S of Moon
15	16:06	LAST QUARTER MOON
17	10:08	Moon at Apogee: 404651 km
17	21	Leonid Meteor Shower
20	14:34	Spica 3.5°S of Moon
21	08:55	Saturn 2.0°N of Moon
23	13	Mercury at Superior Conjunction
23	17:36	NEW MOON
30	19:49	FIRST QUARTER MOON
Dec 02	16:20	Moon at Perigee: 369538 km
06	12:43	Pleiades 2.1°N of Moon
07	17:42	FULL MOON
10	14:08	Pollux 1.7°N of Moon
13	04:15	Mars 2.1°S of Moon
14	10	Geminid Meteor Shower
15	07:09	Moon at Apogee: 404332 km
15	13:32	LAST QUARTER MOON
17	23:15	Spica 3.4°S of Moon
18	23:21	Saturn 2.5°N of Moon
20	09:24	Jupiter 2.6°N of Moon
21	12:19	Winter Solstice
21	16:30	Antares 1.7°S of Moon
22	18	Ursid Meteor Shower
23	08:06	NEW MOON
24	06:17	Mars 3.4°N of Regulus
24	14:14	Mercury 2.8°N of Moon
27	09:54	Moon at Perigee: 368408 km
30	03:46	FIRST QUARTER MOON

Almanac of Astronomical Events for 2042

Date	GMT (h:m)	Event
Jan 01	11	Venus at Superior Conjunction
02	19:59	Pleiades 1.9°N of Moon
02	22	Mercury at Greatest Elong: 19.5°E
04	00	Quadrantid Meteor Shower
04	09	Earth at Perihelion: 0.98330 AU
06	08:54	FULL MOON
06	22:45	Pollux 1.8°N of Moon
09	14:14	Mars 0.7°S of Moon: Occn.
12	04:04	Moon at Apogee: 404808 km
14	07:55	Spica 3.1°S of Moon
14	11:24	LAST QUARTER MOON
15	12:19	Saturn 3.0°N of Moon
17	04:21	Jupiter 3.2°N of Moon
18	02:23	Antares 1.5°S of Moon
19	02	Mercury at Inferior Conjunction
21	20:42	NEW MOON
23	22:31	Moon at Perigee: 362798 km
28	12:48	FIRST QUARTER MOON
30	01:34	Pleiades 1.7°N of Moon
30	13	Uranus at Opposition
Feb 03	05:44	Pollux 1.7°N of Moon
05	01:58	FULL MOON
05	05:36	Mars 0.2°S of Moon: Occn.
06	12	Mars at Opposition
08	21:14	Moon at Apogee: 405668 km
10	15:39	Spica 2.8°S of Moon
11	21:56	Saturn 3.5°N of Moon
12	02	Mercury at Greatest Elong: 25.9°W
13	07:16	LAST QUARTER MOON
13	19:53	Jupiter 3.8°N of Moon
14	11:40	Antares 1.3°S of Moon
20	07:39	NEW MOON
21	05:54	Moon at Perigee: 358434 km
26	07:38	Pleiades 1.4°N of Moon
26	23:29	FIRST QUARTER MOON
Mar 02	11:26	Pollux 1.6°N of Moon
03	18:14	Mars 1.0°S of Moon: Occn.
06	20:10	FULL MOON
08	05:04	Moon at Apogee: 406261 km
09	22:15	Spica 2.6°S of Moon
11	03:37	Saturn 3.8°N of Moon
13	19:10	Antares 1.1°S of Moon
14	23:21	LAST QUARTER MOON
20	11:53	Vernal Equinox
21	17:23	NEW MOON
21	17:40	Moon at Perigee: 356943 km
21	23	Mars at Aphelion: 1.66607 AU
23	02:23	Venus 0.3°S of Moon: Occn.
25	15:53	Pleiades 1.3°N of Moon
28	12:00	FIRST QUARTER MOON
29	13	Mercury at Superior Conjunction
29	17:18	Pollux 1.5°N of Moon
30	22:18	Mars 2.2°S of Moon
Apr 04	05:51	Moon at Apogee: 406302 km
05	14:16	FULL MOON

Date	GMT (h:m)	Event
Apr 05	14:29	Pen. Lunar Eclipse; mag=0.868
06	04:20	Spica 2.6°S of Moon
07	06:26	Saturn 3.8°N of Moon
10	01:09	Antares 1.0°S of Moon
13	11:09	LAST QUARTER MOON
19	04:27	Moon at Perigee: 358558 km
20	02:16	Total Solar Eclipse; mag=1.061
20	02:19	NEW MOON
21	12:09	Mercury 0.4°N of Moon: Occn.
22	01:26	Venus 2.3°S of Moon
22	02:05	Pleiades 1.3°N of Moon
22	21	Lyrid Meteor Shower
23	03:12	Venus 3.3°S of Pleiades
23	07	Saturn at Opposition
24	13	Mercury at Greatest Elong: 20.3°E
26	00:41	Pollux 1.4°N of Moon
27	02:19	FIRST QUARTER MOON
27	18:15	Mars 3.0°S of Moon
May 01	15:06	Moon at Apogee: 405696 km
02	01:38	Mercury 1.7°S of Pleiades
03	10:44	Spica 2.6°S of Moon
04	08:35	Saturn 3.6°N of Moon
05	06:48	FULL MOON
05	11	Eta-Aquarid Meteor Shower
07	06:54	Antares 1.0°S of Moon
12	19:18	LAST QUARTER MOON
16	08	Mercury at Inferior Conjunction
17	09:30	Moon at Perigee: 362606 km
17	21	Jupiter at Opposition
19	10:55	NEW MOON
21	23:49	Venus 3.1°S of Moon
23	09:36	Pollux 1.5°N of Moon
26	00:40	Mars 3.2°S of Moon
26	18:18	FIRST QUARTER MOON
29	07:12	Moon at Apogee: 404716 km
30	07:11	Mars 0.9°N of Regulus
30	17:56	Spica 2.5°S of Moon
31	12:11	Saturn 3.4°N of Moon
Jun 02	11:46	Jupiter 3.8°N of Moon
03	13:36	Antares 1.0°S of Moon
03	20:48	FULL MOON
11	01:00	LAST QUARTER MOON
11	15	Mercury at Greatest Elong: 23.7°W
13	22:24	Moon at Perigee: 367563 km
15	21:38	Pleiades 1.2°N of Moon
17	19:48	NEW MOON
19	18:59	Pollux 1.7°N of Moon
20	23:13	Venus 3.0°S of Moon
21	05:16	Summer Solstice
21	09:53	Mercury 3.2°N of Aldebaran
23	13:44	Mars 2.7°S of Moon
25	11:29	FIRST QUARTER MOON
26	01:31	Moon at Apogee: 404088 km
27	01:53	Spica 2.3°S of Moon
27	18:27	Saturn 3.4°N of Moon
29	14:59	Jupiter 3.6°N of Moon
30	21:37	Antares 1.0°S of Moon

Jul 03	08:09	FULL MOON
06	13	Earth at Aphelion: 1.01666 AU
08	21:10	Venus 0.9°N of Regulus
09	07:47	Moon at Perigee: 369567 km
10	05:38	LAST QUARTER MOON
12	14	Mercury at Superior Conjunction
13	04:36	Pleiades 1.1°N of Moon
17	05:52	NEW MOON
19	20:56	Regulus 3.8°S of Moon
20	23:09	Venus 2.7°S of Moon
22	07:20	Mars 1.7°S of Moon
23	20:11	Moon at Apogee: 404290 km
24	10:04	Spica 2.0°S of Moon
25	03:22	Saturn 3.6°N of Moon
25	05:01	FIRST QUARTER MOON
26	22:31	Jupiter 3.7°N of Moon
28	06:27	Antares 0.8°S of Moon
28	12	Delta-Aquarid Meteor Shower
Aug 01	17:33	FULL MOON
02	03:55	Mercury 0.5°N of Regulus
04	18:09	Moon at Perigee: 365543 km
05	11	Uranus in Conjunction with Sun
08	10:35	LAST QUARTER MOON
09	10:10	Pleiades 0.9°N of Moon
10	09	Venus at Greatest Elong: 45.8°E
13	04	Perseid Meteor Shower
13	10:27	Pollux 1.7°N of Moon
15	18:01	NEW MOON
17	23:39	Mercury 4.1°S of Moon
19	17:07	Venus 3.2°S of Moon
20	04:04	Mars 0.1°S of Moon: Occn.
20	13:59	Moon at Apogee: 405193 km
20	17:47	Spica 1.8°S of Moon
21	14:14	Saturn 3.8°N of Moon
22	22	Mercury at Greatest Elong: 27.4°E
23	10:00	Jupiter 3.9°N of Moon
23	21:55	FIRST QUARTER MOON
24	15:05	Antares 0.6°S of Moon
27	02	Venus 4.0°S of Mars
31	02:02	FULL MOON
31	15:12	Mars 1.9°N of Spica
Sep 01	15:49	Moon at Perigee: 360666 km
02	16:35	Venus 1.7°S of Spica
05	12	Neptune at Perihelion: 29.80642 AU
05	16:04	Pleiades 0.7°N of Moon
06	17:09	LAST QUARTER MOON
09	16:06	Pollux 1.5°N of Moon
12	10:59	Regulus 3.7°S of Moon
14	08:50	NEW MOON
16	03	Mars 2.5°S of Saturn
17	00:39	Spica 1.7°S of Moon
17	04:21	Moon at Apogee: 406154 km
18	02:34	Mars 1.5°N of Moon
18	18	Mercury at Inferior Conjunction
20	22:35	Antares 0.4°S of Moon
22	13:20	FIRST QUARTER MOON
22	21:11	Autumnal Equinox
29	10:34	FULL MOON
29	10:44	Pen. Lunar Eclipse; mag=0.953
29	23:50	Moon at Perigee: 357428 km

Oct 02	23:59	Pleiades 0.6°N of Moon
04	11	Mercury at Greatest Elong: 17.9°W
06	02:35	LAST QUARTER MOON
06	21:47	Pollux 1.4°N of Moon
09	16:41	Regulus 3.8°S of Moon
14	01:59	Annular Solar Eclipse; mag=0.930
14	02:03	NEW MOON
14	10:02	Moon at Apogee: 406527 km
17	01:35	Mars 2.8°N of Moon
18	04:47	Antares 0.4°S of Moon
19	06	Venus at Inferior Conjunction
21	21	Orionid Meteor Shower
22	02:53	FIRST QUARTER MOON
28	11:27	Moon at Perigee: 356973 km
28	19:48	FULL MOON
30	10:15	Pleiades 0.6°N of Moon
30	14	Mars 1.2°S of Jupiter
Nov 01	05	Neptune at Opposition
01	16	Saturn in Conjunction with Sun
03	05:04	Pollux 1.5°N of Moon
03	06	Mercury at Superior Conjunction
04	15:51	LAST QUARTER MOON
05	22:52	Regulus 3.8°S of Moon
06	01:33	Mars 3.8°N of Antares
10	00:46	Venus 2.3°S of Moon
10	12:15	Moon at Apogee: 406245 km
10	13:01	Spica 1.7°S of Moon
12	20:28	NEW MOON
14	10:34	Antares 0.5°S of Moon
15	00:28	Mars 3.7°N of Moon
18	03	Leonid Meteor Shower
20	14:31	FIRST QUARTER MOON
25	21:12	Venus 3.1°N of Spica
25	22:40	Moon at Perigee: 359651 km
26	21:30	Pleiades 0.6°N of Moon
27	06:06	FULL MOON
30	14:32	Pollux 1.6°N of Moon
Dec 03	06:37	Regulus 3.6°S of Moon
04	09:19	LAST QUARTER MOON
04	21	Jupiter in Conjunction with Sun
07	03	Mercury 1.4°S of Mars
07	19:53	Spica 1.5°S of Moon
08	00:37	Moon at Apogee: 405435 km
08	16:10	Venus 4.1°N of Moon
12	14:29	NEW MOON
13	23:23	Mars 4.0°N of Moon
14	07:30	Mercury 3.0°N of Moon
14	16	Geminid Meteor Shower
17	01	Mercury at Greatest Elong: 20.4°E
20	00:28	FIRST QUARTER MOON
21	18:04	Winter Solstice
23	00	Ursid Meteor Shower
23	16	Venus 0.9°N of Saturn
24	02:12	Moon at Perigee: 364795 km
24	07:29	Pleiades 0.5°N of Moon
26	17:43	FULL MOON
28	01:00	Pollux 1.8°N of Moon
29	23	Venus at Greatest Elong: 46.9°W

Almanac of Astronomical Events for 2043

Date	GMT (h:m)	Event
Jan 02	23	Earth at Perihelion: 0.98329 AU
03	06	Mercury at Inferior Conjunction
03	06:08	LAST QUARTER MOON
04	03:42	Spica 1.3°S of Moon
04	06	Quadrantid Meteor Shower
04	19:56	Moon at Apogee: 404558 km
08	00:55	Antares 0.4°S of Moon
11	06:53	NEW MOON
18	09:05	FIRST QUARTER MOON
19	20:02	Moon at Perigee: 369914 km
20	14:49	Pleiades 0.3°N of Moon
24	10:30	Pollux 1.8°N of Moon
25	06:56	FULL MOON
25	11	Mercury at Greatest Elong: 24.6°W
27	01:34	Regulus 3.1°S of Moon
29	12	Venus 2.0°N of Jupiter
31	12:13	Spica 1.0°S of Moon
Feb 01	17:22	Moon at Apogee: 404270 km
02	04:14	LAST QUARTER MOON
04	09:33	Antares 0.2°S of Moon
04	10	Uranus at Opposition
08	02:34	Mercury 4.2°N of Moon
09	21:07	NEW MOON
13	17:40	Moon at Perigee: 367963 km
16	17:00	FIRST QUARTER MOON
16	20:20	Pleiades 0.1°N of Moon
20	17:44	Pollux 1.7°N of Moon
20	18	Mars in Conjunction with Sun
23	09:54	Regulus 3.1°S of Moon
23	21:58	FULL MOON
27	20:29	Spica 0.8°S of Moon
28	10	Mars at Perihelion: 1.38144 AU
Mar 01	13:38	Moon at Apogee: 404776 km
03	17:59	Antares 0.0°N of Moon
04	01:07	LAST QUARTER MOON
11	09:09	NEW MOON
12	21	Mercury at Superior Conjunction
13	09:10	Moon at Perigee: 362515 km
16	02:28	Pleiades 0.0°S of Moon
18	01:03	FIRST QUARTER MOON
19	23:17	Pollux 1.6°N of Moon
20	17:29	Vernal Equinox
22	16:25	Regulus 3.1°S of Moon
25	14:26	FULL MOON
25	14:31	Total Lunar Eclipse; mag=1.114
27	03:50	Spica 0.7°S of Moon
29	05:28	Moon at Apogee: 405650 km
31	01:24	Antares 0.1°N of Moon

Date	GMT (h:m)	Event
Apr 02	18:56	LAST QUARTER MOON
07	06	Mercury at Greatest Elong: 19.2°E
07	11:13	Venus 0.9°N of Moon: Occn.
09	18:56	Total Solar Eclipse; mag=1.010
09	19:06	NEW MOON
10	16:14	Moon at Perigee: 358473 km
11	01:54	Mercury 0.3°N of Moon: Occn.
12	10:55	Pleiades 0.0°S of Moon
16	05:08	Pollux 1.6°N of Moon
16	10:09	FIRST QUARTER MOON
18	22:00	Regulus 3.1°S of Moon
23	04	Lyrid Meteor Shower
23	10:12	Spica 0.8°S of Moon
24	07:23	FULL MOON
25	12:50	Moon at Apogee: 406255 km
26	17	Mercury at Inferior Conjunction
27	07:46	Antares 0.0°N of Moon
30	11	Neptune in Conjunction with Sun
May 02	08:59	LAST QUARTER MOON
05	17	Eta-Aquarid Meteor Shower
05	19	Saturn at Opposition
07	06	Mercury 1.2°S of Mars
07	11:55	Venus 3.2°S of Moon
08	01:48	Mars 2.9°S of Moon
08	01:48	Mercury 4.3°S of Moon
09	02:32	Moon at Perigee: 357136 km
09	03:21	NEW MOON
13	12:52	Pollux 1.8°N of Moon
14	21	Mercury 1.5°S of Venus
15	21:05	FIRST QUARTER MOON
16	04:17	Regulus 3.0°S of Moon
20	16:12	Spica 0.7°S of Moon
22	14:46	Moon at Apogee: 406276 km
23	23:37	FULL MOON
24	05	Mercury at Greatest Elong: 25.3°W
24	13:43	Antares 0.1°S of Moon
25	13	Venus 1.0°S of Mars
31	19:25	LAST QUARTER MOON
Jun 04	14	Mercury 2.3°S of Mars
05	22:11	Mars 4.1°S of Moon
06	08:06	Pleiades 0.0°N of Moon
06	11:59	Moon at Perigee: 358682 km
07	10:35	NEW MOON
09	22:30	Pollux 1.9°N of Moon
12	12:14	Regulus 2.7°S of Moon
14	10:19	FIRST QUARTER MOON
16	22:41	Spica 0.5°S of Moon
18	23:28	Moon at Apogee: 405646 km
20	00	Jupiter at Opposition
20	20:01	Antares 0.0°S of Moon
21	10:59	Summer Solstice
22	14:20	FULL MOON
27	00	Mercury at Superior Conjunction
30	02:53	LAST QUARTER MOON

Date	GMT (h:m)	Event
Jul 03	17:26	Pleiades 0.1°S of Moon
04	15:40	Mars 4.7°S of Moon
04	16:08	Moon at Perigee: 362588 km
06	02	Earth at Aphelion: 1.01674 AU
06	17:51	NEW MOON
09	21:34	Regulus 2.5°S of Moon
14	01:47	FIRST QUARTER MOON
14	06:10	Spica 0.2°S of Moon
16	14:34	Moon at Apogee: 404723 km
18	03:08	Antares 0.1°N of Moon
22	03:24	FULL MOON
27	06:19	Mercury 0.3°S of Regulus
28	19	Delta-Aquarid Meteor Shower
29	08:23	LAST QUARTER MOON
31	00:32	Pleiades 0.3°S of Moon
Aug 01	05:13	Moon at Perigee: 367517 km
02	06:46	Mars 4.7°S of Moon
03	17:56	Pollux 2.0°N of Moon
05	02:23	NEW MOON
05	06	Mercury at Greatest Elong: 27.3°E
06	07:07	Regulus 2.4°S of Moon
07	03:24	Mercury 3.9°S of Moon
07	12	Venus at Superior Conjunction
10	14:28	Spica 0.0°N of Moon
10	15	Uranus in Conjunction with Sun
12	18:57	FIRST QUARTER MOON
13	08:32	Moon at Apogee: 404190 km
13	10	Perseid Meteor Shower
14	11:01	Antares 0.3°N of Moon
20	15:04	FULL MOON
26	13:38	Moon at Perigee: 369598 km
27	06:05	Pleiades 0.5°S of Moon
27	13:09	LAST QUARTER MOON
30	20:16	Mars 4.2°S of Moon
31	01:14	Pollux 1.9°N of Moon
Sep 01	19	Mercury at Inferior Conjunction
03	13:17	NEW MOON
06	22:55	Spica 0.1°N of Moon
10	03:40	Moon at Apogee: 404492 km
10	19:07	Antares 0.4°N of Moon
11	13:01	FIRST QUARTER MOON
18	02	Mercury at Greatest Elong: 17.9°W
19	01:47	FULL MOON
19	01:50	Total Lunar Eclipse; mag=1.256
22	00:24	Moon at Perigee: 365389 km
23	03:07	Autumnal Equinox
23	11:58	Pleiades 0.5°S of Moon
25	18:40	LAST QUARTER MOON
27	06:52	Pollux 1.9°N of Moon
28	08:49	Mars 3.1°S of Moon
29	22:19	Regulus 2.4°S of Moon

Date	GMT (h:m)	Event
Oct 02	04:31	Venus 2.6°N of Spica
03	03:00	Annular Solar Eclipse; mag=0.950
03	03:12	NEW MOON
04	15:08	Venus 3.5°N of Moon
07	22:23	Moon at Apogee: 405459 km
08	02:44	Antares 0.4°N of Moon
11	07:05	FIRST QUARTER MOON
15	03	Mercury at Superior Conjunction
18	11:56	FULL MOON
19	23:27	Moon at Perigee: 360242 km
20	20:01	Pleiades 0.5°S of Moon
22	00	Venus 2.0°S of Saturn
22	03	Orionid Meteor Shower
24	12:34	Pollux 2.0°N of Moon
25	02:27	LAST QUARTER MOON
26	20:45	Mars 1.5°S of Moon
27	03:54	Regulus 2.3°S of Moon
Nov 01	19:57	NEW MOON
03	10:38	Mars 1.1°N of Regulus
03	17	Neptune at Opposition
04	00:41	Venus 4.2°N of Moon
04	09:30	Antares 0.3°N of Moon
04	13:03	Moon at Apogee: 406391 km
08	05:43	Venus 3.8°N of Antares
10	00:13	FIRST QUARTER MOON
13	17	Saturn in Conjunction with Sun
15	00:12	Mercury 2.4°N of Antares
16	21:52	FULL MOON
17	06:35	Pleiades 0.4°S of Moon
17	09:11	Moon at Perigee: 356949 km
18	09	Leonid Meteor Shower
20	20:17	Pollux 2.2°N of Moon
23	10:02	Regulus 2.1°S of Moon
23	13:46	LAST QUARTER MOON
24	07:28	Mars 0.6°N of Moon: Occn.
27	19:13	Spica 0.2°N of Moon
29	21	Venus 1.5°S of Jupiter
29	22	Mercury at Greatest Elong: 21.6°E
Dec 01	14:37	NEW MOON
01	17:07	Moon at Apogee: 406706 km
03	12:57	Mercury 3.0°N of Moon
04	08:27	Venus 3.2°N of Moon
08	00	Mercury 4.3°S of Jupiter
09	15:27	FIRST QUARTER MOON
14	18:03	Pleiades 0.4°S of Moon
14	22	Geminid Meteor Shower
15	22:01	Moon at Perigee: 356769 km
16	08:02	FULL MOON
18	06:35	Pollux 2.3°N of Moon
18	13	Mercury at Inferior Conjunction
20	18:20	Regulus 1.8°S of Moon
22	00:02	Winter Solstice
22	14:50	Mars 2.7°N of Moon
23	05:04	LAST QUARTER MOON
23	07	Ursid Meteor Shower
25	01:30	Spica 0.5°N of Moon
28	19:33	Moon at Apogee: 406403 km
28	21:51	Antares 0.3°N of Moon
31	09:48	NEW MOON

Almanac of Astronomical Events for 2044

Date	GMT (h:m)	Event
Jan 03	15:31	Venus 1.0°N of Moon: Occn.
04	12	Quadrantid Meteor Shower
05	14	Earth at Perihelion: 0.98329 AU
05	18	Jupiter in Conjunction with Sun
07	20	Mercury at Greatest Elong: 23.1°W
08	04:02	FIRST QUARTER MOON
11	04:04	Pleiades 0.6°S of Moon
13	09:08	Moon at Perigee: 359906 km
14	17:54	Pollux 2.4°N of Moon
14	18:51	FULL MOON
17	04:42	Regulus 1.6°S of Moon
21	09:10	Spica 0.8°N of Moon
21	23:47	LAST QUARTER MOON
25	04:44	Antares 0.5°N of Moon
25	09:32	Moon at Apogee: 405544 km
27	17	Mercury 0.8°S of Jupiter
28	12:41	Jupiter 4.4°N of Moon
28	15:10	Mercury 3.4°N of Moon
30	04:04	NEW MOON
Feb 02	18:44	Venus 1.2°S of Moon
06	13:46	FIRST QUARTER MOON
07	00	Mars at Aphelion: 1.66600 AU
07	11:16	Pleiades 0.8°S of Moon
09	08	Uranus at Opposition
10	10:42	Moon at Perigee: 365312 km
11	03:54	Pollux 2.4°N of Moon
13	06:42	FULL MOON
13	15:21	Regulus 1.5°S of Moon
17	18:12	Spica 0.9°N of Moon
20	20:20	LAST QUARTER MOON
21	12:30	Antares 0.6°N of Moon
22	05:18	Moon at Apogee: 404628 km
23	14	Mercury at Superior Conjunction
25	08:12	Jupiter 4.2°N of Moon
28	20:12	NEW MOON
28	20:23	Annular Solar Eclipse; mag=0.960
Mar 03	12:58	Venus 1.9°S of Moon
05	16:43	Pleiades 0.9°S of Moon
06	21:17	FIRST QUARTER MOON
07	20:36	Moon at Perigee: 369930 km
09	11:14	Pollux 2.3°N of Moon
11	12	Mars at Opposition
12	00:19	Regulus 1.5°S of Moon
13	17:11	Mars 4.4°N of Moon
13	19:37	Total Lunar Eclipse; mag=1.203
13	19:41	FULL MOON
16	03:29	Spica 1.0°N of Moon
17	20	Venus at Greatest Elong: 46.2°E
19	20:46	Antares 0.6°N of Moon
19	23:20	Vernal Equinox
20	09	Mercury at Greatest Elong: 18.5°E
21	01:56	Moon at Apogee: 404337 km
21	16:52	LAST QUARTER MOON
24	02:46	Jupiter 3.9°N of Moon
29	09:26	NEW MOON

Date	GMT (h:m)	Event
Apr 01	19:28	Venus 0.8°S of Moon: Occn.
01	22:50	Pleiades 0.9°S of Moon
02	02:16	Moon at Perigee: 367395 km
04	05:23	Venus 0.1°N of Pleiades
05	03:45	FIRST QUARTER MOON
05	16:46	Pollux 2.4°N of Moon
06	22	Mercury at Inferior Conjunction
08	06:57	Regulus 1.5°S of Moon
09	08:29	Mars 2.6°N of Moon
12	09:39	FULL MOON
12	11:39	Spica 0.9°N of Moon
16	04:43	Antares 0.5°N of Moon
17	20:46	Moon at Apogee: 404858 km
20	11:48	LAST QUARTER MOON
20	18:29	Jupiter 3.5°N of Moon
22	10	Lyrid Meteor Shower
26	03:12	Mercury 4.6°S of Moon
27	19:42	NEW MOON
29	07:11	Pleiades 0.8°S of Moon
29	18:30	Moon at Perigee: 362200 km
30	06:41	Venus 0.0°N of Moon: Occn.
May 02	00	Neptune in Conjunction with Sun
02	22:41	Pollux 2.5°N of Moon
04	10:28	FIRST QUARTER MOON
04	23	Mercury at Greatest Elong: 26.7°W
04	23	Eta-Aquarid Meteor Shower
05	12:28	Regulus 1.3°S of Moon
06	13:43	Mars 1.5°N of Moon
09	18:16	Spica 1.0°N of Moon
12	00:16	FULL MOON
13	11:48	Antares 0.4°N of Moon
15	11:19	Moon at Apogee: 405736 km
17	04	Saturn at Opposition
18	05:25	Jupiter 3.0°N of Moon
20	04:02	LAST QUARTER MOON
27	03:39	NEW MOON
27	19	Venus at Inferior Conjunction
28	00:15	Moon at Perigee: 358408 km
30	06:40	Pollux 2.7°N of Moon
Jun 01	18:48	Regulus 1.0°S of Moon
02	18:33	FIRST QUARTER MOON
03	09:41	Mars 1.5°N of Moon
06	00:01	Spica 1.1°N of Moon
09	18:01	Antares 0.4°N of Moon
10	11	Mercury at Superior Conjunction
10	15:16	FULL MOON
11	18:31	Moon at Apogee: 406330 km
14	10:41	Jupiter 2.8°N of Moon
18	17:00	LAST QUARTER MOON
20	16:50	Summer Solstice
23	03:57	Pleiades 0.8°S of Moon
25	09:26	Moon at Perigee: 357175 km
25	10:24	NEW MOON
26	15:22	Mercury 2.0°S of Moon
26	16:39	Pollux 2.9°N of Moon
29	03:07	Regulus 0.8°S of Moon

Jul 01	16:03	Mars 2.3°N of Moon
02	04:48	FIRST QUARTER MOON
03	06:08	Spica 1.4°N of Moon
03	16	Earth at Aphelion: 1.01670 AU
06	23:57	Antares 0.5°N of Moon
08	20:57	Moon at Apogee: 406301 km
10	06:22	FULL MOON
11	11:22	Jupiter 2.7°N of Moon
14	21:44	Venus 1.2°N of Aldebaran
17	09	Mercury at Greatest Elong: 26.7°E
18	02:47	LAST QUARTER MOON
20	13:12	Pleiades 1.0°S of Moon
23	18:16	Moon at Perigee: 358737 km
24	06	Jupiter at Opposition
24	17:10	NEW MOON
26	02:18	Mercury 2.3°S of Regulus
26	08:42	Mercury 4.2°S of Moon
26	13:09	Regulus 0.6°S of Moon
28	01	Delta-Aquarid Meteor Shower
30	05:21	Mars 3.2°N of Moon
30	13:38	Spica 1.6°N of Moon
31	17:40	FIRST QUARTER MOON
Aug 03	06:21	Antares 0.6°N of Moon
05	05:22	Moon at Apogee: 405646 km
06	01	Venus at Greatest Elong: 45.8°W
07	10:50	Jupiter 2.9°N of Moon
07	19:49	Mars 1.5°N of Spica
08	21:14	FULL MOON
12	17	Perseid Meteor Shower
14	06	Mercury at Inferior Conjunction
14	19	Uranus in Conjunction with Sun
16	10:03	LAST QUARTER MOON
16	20:15	Pleiades 1.1°S of Moon
20	13:00	Pollux 2.9°N of Moon
20	23:01	Moon at Perigee: 362698 km
23	01:06	NEW MOON
23	01:16	Total Solar Eclipse; mag=1.036
26	22:33	Spica 1.7°N of Moon
27	23:19	Mars 4.0°N of Moon
30	09:18	FIRST QUARTER MOON
30	13:43	Antares 0.7°N of Moon
31	14	Mercury at Greatest Elong: 18.2°W
Sep 01	20:26	Moon at Apogee: 404738 km
03	12:56	Jupiter 3.1°N of Moon
07	11:19	Total Lunar Eclipse; mag=1.046
07	11:24	FULL MOON
07	11:56	Mercury 0.7°N of Regulus
13	01:46	Pleiades 1.1°S of Moon
14	15:58	LAST QUARTER MOON
16	20:31	Pollux 2.9°N of Moon
17	12:14	Moon at Perigee: 367771 km
18	05:59	Venus 2.9°S of Moon
19	08:39	Regulus 0.6°S of Moon
21	11:03	NEW MOON
22	08:47	Autumnal Equinox
23	08:00	Spica 1.7°N of Moon
25	20:27	Mars 4.3°N of Moon
26	01	Mercury at Superior Conjunction
26	21	Mars 2.8°S of Saturn
26	21:56	Antares 0.7°N of Moon
29	03:30	FIRST QUARTER MOON
29	15:06	Moon at Apogee: 404238 km

Sep 30	20:00	Jupiter 3.1°N of Moon
Oct 01	09:21	Venus 0.0°S of Regulus
07	00:30	FULL MOON
10	07:36	Pleiades 1.0°S of Moon
12	15:35	Moon at Perigee: 369711 km
13	21:52	LAST QUARTER MOON
14	02:12	Pollux 3.0°N of Moon
15	16:53	Mars 3.5°N of Antares
16	15:39	Regulus 0.5°S of Moon
18	04:25	Venus 2.4°N of Moon
20	23:36	NEW MOON
21	09	Orionid Meteor Shower
22	11:20	Mercury 3.4°N of Moon
24	06:20	Antares 0.5°N of Moon
24	20:02	Mars 4.0°N of Moon
27	11:18	Moon at Apogee: 404571 km
28	08:05	Jupiter 2.8°N of Moon
28	23:28	FIRST QUARTER MOON
Nov 01	00	Mercury 4.1°S of Saturn
05	05	Neptune at Opposition
05	12:27	FULL MOON
06	15:36	Pleiades 0.9°S of Moon
08	06:14	Moon at Perigee: 364997 km
08	23:11	Mercury 1.8°N of Antares
10	07:57	Pollux 3.2°N of Moon
11	15	Mercury at Greatest Elong: 22.9°E
12	05:09	LAST QUARTER MOON
12	21:13	Regulus 0.3°S of Moon
15	01:06	Venus 3.5°N of Spica
16	23:46	Spica 1.7°N of Moon
17	15	Leonid Meteor Shower
19	14:58	NEW MOON
21	07:46	Mercury 3.0°N of Moon
22	21:55	Mars 3.1°N of Moon
24	06:42	Moon at Apogee: 405513 km
24	11	Saturn in Conjunction with Sun
24	23:50	Jupiter 2.2°N of Moon
27	19:36	FIRST QUARTER MOON
Dec 01	20	Mercury at Inferior Conjunction
04	01:59	Pleiades 0.9°S of Moon
04	23:34	FULL MOON
06	08:05	Moon at Perigee: 359749 km
07	15:58	Pollux 3.4°N of Moon
10	03:28	Regulus 0.0°N of Moon
11	14:52	LAST QUARTER MOON
14	04	Geminid Meteor Shower
14	05:30	Spica 1.9°N of Moon
17	20:37	Antares 0.4°N of Moon
18	10	Mercury 0.6°N of Saturn
18	15	Venus 0.7°S of Saturn
19	08:53	NEW MOON
20	09	Mercury at Greatest Elong: 21.7°W
21	05:43	Winter Solstice
21	20:29	Moon at Apogee: 406354 km
22	02:06	Mars 1.6°N of Moon
22	13	Ursid Meteor Shower
22	17:46	Jupiter 1.6°N of Moon
27	14:00	FIRST QUARTER MOON
31	13:05	Pleiades 1.0°S of Moon

Almanac of Astronomical Events for 2045

Date	GMT (h:m)	Event
Jan 03	03	Mercury 0.1°S of Venus
03	10:20	FULL MOON
03	15	Earth at Perihelion: 0.98326 AU
03	19	Quadrantid Meteor Shower
03	19:25	Moon at Perigee: 356773 km
04	02:40	Pollux 3.5°N of Moon
04	13	Mars 0.5°S of Jupiter
06	12:16	Regulus 0.2°N of Moon
10	03:32	LAST QUARTER MOON
10	11:34	Spica 2.2°N of Moon
14	02:28	Antares 0.6°N of Moon
15	08	Mars at Perihelion: 1.38130 AU
17	22:35	Moon at Apogee: 406609 km
18	04:25	NEW MOON
20	08:00	Mars 0.4°S of Moon: Occn.
26	05:09	FIRST QUARTER MOON
27	22:36	Pleiades 1.2°S of Moon
31	14:15	Pollux 3.5°N of Moon
Feb 01	08:43	Moon at Perigee: 357106 km
01	21:05	FULL MOON
02	23:20	Regulus 0.3°N of Moon
04	12	Mercury at Superior Conjunction
06	19:33	Spica 2.3°N of Moon
07	15	Jupiter in Conjunction with Sun
08	19:03	LAST QUARTER MOON
10	08:43	Antares 0.7°N of Moon
13	06	Uranus at Opposition
14	02:57	Moon at Apogee: 406254 km
16	23:51	NEW MOON
16	23:55	Annular Solar Eclipse; mag=0.928
18	13:49	Mars 2.5°S of Moon
23	20	Mercury 0.7°N of Mars
24	05:30	Pleiades 1.2°S of Moon
24	16:37	FIRST QUARTER MOON
28	00:18	Pollux 3.5°N of Moon
Mar 01	18:43	Moon at Perigee: 360563 km
02	10:32	Regulus 0.3°N of Moon
03	07:42	Pen. Lunar Eclipse; mag=0.962
03	07:52	FULL MOON
03	18	Mercury at Greatest Elong: 18.2°E
06	05:26	Spica 2.3°N of Moon
09	16:21	Antares 0.7°N of Moon
10	12:50	LAST QUARTER MOON
13	18:23	Moon at Apogee: 405311 km
16	05:24	Jupiter 0.4°S of Moon: Occn.
18	17:15	NEW MOON
18	22	Venus at Superior Conjunction
20	01	Mercury at Inferior Conjunction
20	05:08	Vernal Equinox
23	10:54	Pleiades 1.1°S of Moon
26	00:56	FIRST QUARTER MOON
27	07:36	Pollux 3.6°N of Moon
29	17:25	Moon at Perigee: 365846 km
29	19:48	Regulus 0.4°N of Moon

Date	GMT (h:m)	Event
Apr 01	18:43	FULL MOON
02	15:39	Spica 2.3°N of Moon
06	01:11	Antares 0.5°N of Moon
09	07:52	LAST QUARTER MOON
10	13:48	Moon at Apogee: 404371 km
13	01:32	Jupiter 1.2°S of Moon
16	22	Mercury at Greatest Elong: 27.6°W
17	07:27	NEW MOON
19	16:58	Pleiades 1.0°S of Moon
22	16	Lyrid Meteor Shower
23	13:08	Pollux 3.8°N of Moon
24	07:12	FIRST QUARTER MOON
24	22:26	Moon at Perigee: 369865 km
26	02:32	Regulus 0.5°N of Moon
30	00:30	Spica 2.3°N of Moon
May 01	05:52	FULL MOON
02	10	Mars in Conjunction with Sun
03	10:07	Antares 0.4°N of Moon
04	14	Neptune in Conjunction with Sun
05	05	Eta-Aquarid Meteor Shower
08	09:24	Moon at Apogee: 404121 km
09	02:51	LAST QUARTER MOON
10	19:54	Jupiter 2.0°S of Moon
16	18:26	NEW MOON
17	23:39	Venus 4.4°S of Moon
20	10:22	Moon at Perigee: 367210 km
20	19:05	Pollux 4.0°N of Moon
23	08:03	Regulus 0.8°N of Moon
23	12:38	FIRST QUARTER MOON
25	23	Mercury at Superior Conjunction
27	07:21	Spica 2.4°N of Moon
29	10	Saturn at Opposition
30	17:52	FULL MOON
30	18:01	Antares 0.3°N of Moon
Jun 05	03:25	Moon at Apogee: 404694 km
07	10:37	Jupiter 2.8°S of Moon
07	20:23	LAST QUARTER MOON
13	10:33	Pleiades 1.0°S of Moon
15	03:05	NEW MOON
16	16:30	Mercury 1.2°S of Moon: Occn.
16	19:10	Venus 1.5°S of Moon
17	02:09	Moon at Perigee: 362268 km
17	03:01	Pollux 4.2°N of Moon
19	14:25	Regulus 1.1°N of Moon
20	22:34	Summer Solstice
21	18:28	FIRST QUARTER MOON
22	09	Mercury 0.4°S of Venus
23	12:59	Spica 2.6°N of Moon
27	00:32	Antares 0.4°N of Moon
29	05	Mercury at Greatest Elong: 25.6°E
29	07:16	FULL MOON

Date	GMT (h:m)	Event
Jul 02	18:04	Moon at Apogee: 405608 km
04	19:56	Jupiter 3.2°S of Moon
06	13	Earth at Aphelion: 1.01669 AU
07	11:31	LAST QUARTER MOON
10	20:30	Pleiades 1.1°S of Moon
13	03:07	Mars 3.4°S of Moon
14	10:28	NEW MOON
15	07:12	Moon at Perigee: 358581 km
15	12:59	Mercury 4.9°S of Moon
16	16:10	Venus 2.0°N of Moon
16	22:52	Regulus 1.2°N of Moon
20	09:08	Venus 1.0°N of Regulus
20	18:58	Spica 2.8°N of Moon
21	01:52	FIRST QUARTER MOON
24	06:12	Antares 0.5°N of Moon
26	22	Mercury at Inferior Conjunction
28	07	Delta-Aquarid Meteor Shower
28	22:11	FULL MOON
30	02:00	Moon at Apogee: 406198 km
31	23:19	Jupiter 3.3°S of Moon
Aug 05	23:57	LAST QUARTER MOON
07	05:18	Pleiades 1.2°S of Moon
10	20:01	Mars 1.9°S of Moon
10	23:41	Pollux 4.2°N of Moon
11	09:35	Mercury 4.2°S of Moon
12	16:40	Moon at Perigee: 357292 km
12	17:39	NEW MOON
12	17:41	Total Solar Eclipse; mag=1.077
12	23	Perseid Meteor Shower
14	19	Mercury at Greatest Elong: 18.8°W
15	13:17	Venus 4.4°N of Moon
17	02:40	Spica 2.8°N of Moon
19	11:55	FIRST QUARTER MOON
19	23	Uranus in Conjunction with Sun
20	12:13	Antares 0.5°N of Moon
26	04:13	Moon at Apogee: 406140 km
27	13:53	Pen. Lunar Eclipse; mag=0.683
27	14:08	FULL MOON
27	22:40	Jupiter 3.0°S of Moon
30	09	Jupiter at Opposition
Sep 03	12:09	Pleiades 1.2°S of Moon
04	08:52	Venus 1.3°N of Spica
04	10:03	LAST QUARTER MOON
07	09:25	Pollux 4.2°N of Moon
08	12:14	Mars 0.0°S of Moon: Occn.
08	20	Mercury at Superior Conjunction
09	20:02	Regulus 1.2°N of Moon
10	02:20	Moon at Perigee: 358804 km
11	01:27	NEW MOON
13	12:16	Spica 2.8°N of Moon
14	08:12	Venus 4.0°N of Moon
16	19:36	Antares 0.4°N of Moon
18	01:30	FIRST QUARTER MOON
22	12:20	Moon at Apogee: 405498 km
22	14:33	Autumnal Equinox
23	22:05	Jupiter 2.7°S of Moon
26	06:11	FULL MOON
30	17:09	Mercury 1.4°N of Spica
30	17:38	Pleiades 1.0°S of Moon

Date	GMT (h:m)	Event
Oct 03	18:31	LAST QUARTER MOON
04	16:58	Pollux 4.4°N of Moon
07	03:19	Mars 2.1°N of Moon
07	05:33	Regulus 1.3°N of Moon
08	08:16	Moon at Perigee: 362867 km
09	18:57	Mars 0.8°N of Regulus
10	10:37	NEW MOON
12	02:48	Mercury 3.3°N of Moon
14	01:17	Venus 1.6°N of Moon
14	04:33	Antares 0.2°N of Moon
14	13	Venus 4.9°S of Saturn
16	01:42	Venus 1.2°N of Antares
17	18:55	FIRST QUARTER MOON
20	04:05	Moon at Apogee: 404638 km
21	01:36	Jupiter 2.6°S of Moon
21	15	Orionid Meteor Shower
22	15	Venus at Greatest Elong: 46.9°E
25	04	Mercury at Greatest Elong: 24.2°E
25	21:31	FULL MOON
27	23:29	Pleiades 0.9°S of Moon
Nov 02	02:09	LAST QUARTER MOON
03	12:42	Regulus 1.6°N of Moon
04	16:29	Mars 4.2°N of Moon
04	21:44	Moon at Perigee: 368185 km
07	08:12	Spica 2.7°N of Moon
07	17	Neptune at Opposition
08	21:49	NEW MOON
10	14:09	Antares 0.1°N of Moon
12	13:38	Venus 0.6°S of Moon: Occn.
16	03	Mercury at Inferior Conjunction
16	15:26	FIRST QUARTER MOON
16	23:59	Moon at Apogee: 404209 km
17	11:00	Jupiter 2.9°S of Moon
17	21	Leonid Meteor Shower
24	07:15	Pleiades 0.8°S of Moon
24	11:43	FULL MOON
29	18:10	Moon at Perigee: 369747 km
30	18:15	Regulus 1.9°N of Moon
Dec 01	09:46	LAST QUARTER MOON
03	06	Mercury at Greatest Elong: 20.4°W
04	15:32	Spica 2.9°N of Moon
06	00	Saturn in Conjunction with Sun
08	11:41	NEW MOON
11	00:17	Venus 0.9°N of Moon: Occn.
14	10	Geminid Meteor Shower
14	21:16	Moon at Apogee: 404595 km
15	01:35	Jupiter 3.4°S of Moon
16	13:08	FIRST QUARTER MOON
21	11:36	Winter Solstice
21	16:58	Pleiades 0.9°S of Moon
22	19	Ursid Meteor Shower
24	00:49	FULL MOON
24	21	Mars at Aphelion: 1.66596 AU
26	14:47	Moon at Perigee: 364579 km
28	00:36	Regulus 2.1°N of Moon
30	18:11	LAST QUARTER MOON
31	21:10	Spica 3.1°N of Moon

Almanac of Astronomical Events for 2046

Date	GMT (h:m)	Event
Jan 01	14	Venus at Inferior Conjunction
03	01	Earth at Perihelion: 0.98334 AU
04	01	Quadrantid Meteor Shower
04	05:46	Antares 0.1°N of Moon
07	04:24	NEW MOON
11	16:45	Moon at Apogee: 405530 km
11	19:29	Jupiter 3.9°S of Moon
15	09:42	FIRST QUARTER MOON
16	08	Mercury at Superior Conjunction
18	03:07	Pleiades 1.0°S of Moon
22	12:51	FULL MOON
22	13:01	Partial Lunar Eclipse; mag=0.053
23	19:02	Moon at Perigee: 359440 km
24	09:34	Regulus 2.2°N of Moon
28	03:12	Spica 3.2°N of Moon
29	04:11	LAST QUARTER MOON
31	11:20	Antares 0.2°N of Moon
Feb 05	23:05	Annular Solar Eclipse; mag=0.923
05	23:10	NEW MOON
07	11:36	Mercury 2.9°S of Moon
08	05:11	Moon at Apogee: 406331 km
08	14:55	Jupiter 4.4°S of Moon
14	03:20	FIRST QUARTER MOON
14	11:48	Pleiades 1.0°S of Moon
15	06	Mercury at Greatest Elong: 18.1°E
18	04	Mercury 3.4°N of Jupiter
18	05	Uranus at Opposition
20	20:43	Regulus 2.2°N of Moon
20	23:44	FULL MOON
21	06:43	Moon at Perigee: 356803 km
24	11:33	Spica 3.2°N of Moon
27	16:23	LAST QUARTER MOON
27	17:23	Antares 0.1°N of Moon
Mar 02	20	Mercury at Inferior Conjunction
07	06:49	Moon at Apogee: 406576 km
07	18:15	NEW MOON
13	01	Venus at Greatest Elong: 46.7°W
13	18:23	Pleiades 0.9°S of Moon
15	17:13	FIRST QUARTER MOON
15	18	Jupiter in Conjunction with Sun
20	07:56	Regulus 2.2°N of Moon
20	10:58	Vernal Equinox
21	18:59	Moon at Perigee: 357400 km
22	09:27	FULL MOON
23	22:03	Spica 3.0°N of Moon
27	01:24	Antares 0.0°S of Moon
29	06:57	LAST QUARTER MOON
30	03	Mercury at Greatest Elong: 27.8°W

Date	GMT (h:m)	Event
Apr 02	08:14	Venus 1.0°S of Moon: Occn.
03	12:45	Moon at Apogee: 406176 km
06	11:52	NEW MOON
09	23:52	Pleiades 0.7°S of Moon
14	03:21	FIRST QUARTER MOON
15	06	Mercury 1.5°S of Jupiter
16	17:10	Regulus 2.4°N of Moon
17	16	Mars at Opposition
19	02:59	Moon at Perigee: 360864 km
20	08:51	Spica 3.0°N of Moon
20	18:21	FULL MOON
22	22	Lyrid Meteor Shower
23	11:06	Antares 0.2°S of Moon
27	23:30	LAST QUARTER MOON
29	17:27	Mars 3.3°N of Spica
May 01	03:41	Moon at Apogee: 405206 km
05	11	Eta-Aquarid Meteor Shower
06	02:56	NEW MOON
07	03	Neptune in Conjunction with Sun
08	04	Venus 0.5°S of Jupiter
10	08	Mercury at Superior Conjunction
13	10:25	FIRST QUARTER MOON
13	23:55	Regulus 2.6°N of Moon
16	23:31	Moon at Perigee: 365910 km
17	18:08	Spica 3.1°N of Moon
20	03:15	FULL MOON
20	21:01	Antares 0.3°S of Moon
27	17:06	LAST QUARTER MOON
28	21:55	Moon at Apogee: 404334 km
Jun 04	15:22	NEW MOON
06	12:33	Mercury 0.4°S of Moon: Occn.
10	05:25	Regulus 2.9°N of Moon
10	15	Saturn at Opposition
10	19	Mercury at Greatest Elong: 24.1°E
11	15:27	FIRST QUARTER MOON
12	03:54	Moon at Perigee: 369663 km
14	01:09	Spica 3.2°N of Moon
17	05:39	Antares 0.3°S of Moon
18	13:10	FULL MOON
21	04:15	Summer Solstice
25	16:40	Moon at Apogee: 404192 km
26	10:40	LAST QUARTER MOON
29	17:09	Mars 1.2°N of Spica
30	22:02	Pleiades 0.7°S of Moon

Date	GMT (h:m)	Event
Jul 04	01:39	NEW MOON
05	06	Earth at Aphelion: 1.01673 AU
07	11:40	Regulus 3.0°N of Moon
07	18:06	Moon at Perigee: 367114 km
07	18	Mercury at Inferior Conjunction
10	19:53	FIRST QUARTER MOON
11	06:41	Spica 3.4°N of Moon
11	15:10	Mars 4.5°N of Moon
14	12:24	Antares 0.3°S of Moon
18	00:55	FULL MOON
18	01:05	Partial Lunar Eclipse; mag=0.246
23	10:37	Moon at Apogee: 404859 km
26	03:19	LAST QUARTER MOON
28	07:04	Pleiades 0.7°S of Moon
28	13	Delta-Aquarid Meteor Shower
28	15	Mercury at Greatest Elong: 19.8°W
31	22:00	Venus 1.4°S of Moon
31	23:16	Mercury 2.6°S of Moon
Aug 02	10:20	Total Solar Eclipse; mag=1.053
02	10:25	NEW MOON
02	21	Mercury 1.4°S of Venus
03	19:59	Regulus 3.0°N of Moon
04	09:16	Moon at Perigee: 362170 km
07	12:35	Spica 3.4°N of Moon
08	19:49	Mars 4.0°N of Moon
09	01:15	FIRST QUARTER MOON
10	17:55	Antares 0.3°S of Moon
13	05	Perseid Meteor Shower
16	14:50	FULL MOON
20	01:53	Moon at Apogee: 405828 km
23	06	Mercury at Superior Conjunction
24	15:14	Pleiades 0.6°S of Moon
24	18:36	LAST QUARTER MOON
25	03	Uranus in Conjunction with Sun
31	18:25	NEW MOON
Sep 01	14:28	Moon at Perigee: 358321 km
03	20:26	Spica 3.3°N of Moon
06	06:55	Mars 3.1°N of Moon
06	23:47	Antares 0.4°S of Moon
07	09:07	FIRST QUARTER MOON
15	06:39	FULL MOON
16	10:11	Moon at Apogee: 406400 km
20	21:54	Pleiades 0.5°S of Moon
21	23:56	Mars 2.9°N of Antares
22	20:22	Autumnal Equinox
23	08:16	LAST QUARTER MOON
24	07:26	Mercury 0.6°N of Spica
27	17:03	Regulus 3.1°N of Moon
30	00:47	Moon at Perigee: 356939 km
30	02:25	NEW MOON

Date	GMT (h:m)	Event
Oct 01	06:32	Spica 3.2°N of Moon
01	21:45	Mercury 2.9°N of Moon
04	03	Mars 2.9°S of Saturn
04	07:32	Antares 0.6°S of Moon
04	23:01	Saturn 4.7°N of Moon
04	23:33	Mars 1.8°N of Moon
06	20:41	FIRST QUARTER MOON
07	05	Jupiter at Opposition
07	16	Mercury at Greatest Elong: 25.5°E
13	11:22	Moon at Apogee: 406316 km
14	23:41	FULL MOON
15	16	Venus at Superior Conjunction
18	03:34	Pleiades 0.3°S of Moon
21	21	Orionid Meteor Shower
22	20:07	LAST QUARTER MOON
25	02:34	Regulus 3.3°N of Moon
28	11:41	Moon at Perigee: 358601 km
29	11:17	NEW MOON
31	07	Mercury at Inferior Conjunction
31	17:26	Antares 0.8°S of Moon
Nov 01	12	Jupiter at Perihelion: 4.95342 AU
01	12:17	Saturn 4.3°N of Moon
02	21:31	Mars 0.0°N of Moon: Occn.
05	12:28	FIRST QUARTER MOON
09	19:38	Moon at Apogee: 405682 km
10	04	Neptune at Opposition
13	17:04	FULL MOON
14	09:32	Pleiades 0.2°S of Moon
16	11	Mercury at Greatest Elong: 19.4°W
18	03	Leonid Meteor Shower
21	06:10	LAST QUARTER MOON
21	09:38	Regulus 3.6°N of Moon
25	03:34	Spica 3.3°N of Moon
25	18:09	Moon at Perigee: 363050 km
27	21:50	NEW MOON
29	03:43	Saturn 3.9°N of Moon
Dec 02	00:20	Mars 2.0°S of Moon
03	11	Mars at Perihelion: 1.38138 AU
05	07:56	FIRST QUARTER MOON
07	12:36	Moon at Apogee: 404833 km
11	16:52	Pleiades 0.2°S of Moon
13	09:55	FULL MOON
14	17	Geminid Meteor Shower
17	11	Saturn in Conjunction with Sun
18	15:09	Regulus 3.8°N of Moon
20	14:43	LAST QUARTER MOON
21	17:28	Winter Solstice
22	10:57	Spica 3.4°N of Moon
23	01	Ursid Meteor Shower
23	05:06	Moon at Perigee: 368619 km
25	13:38	Antares 0.8°S of Moon
27	02	Mercury at Superior Conjunction
27	10:39	NEW MOON
28	22:13	Venus 1.5°S of Moon
31	06:26	Mars 4.0°S of Moon

Almanac of Astronomical Events for 2047

Date	GMT (h:m)	Event
Jan 04	05:31	FIRST QUARTER MOON
04	07	Quadrantid Meteor Shower
04	09:26	Moon at Apogee: 404416 km
05	11	Earth at Perihelion: 0.98332 AU
08	01:32	Pleiades 0.3°S of Moon
12	01:21	FULL MOON
12	01:25	Total Lunar Eclipse; mag=1.234
14	21:28	Regulus 3.8°N of Moon
16	21:10	Moon at Perigee: 369325 km
18	16:26	Spica 3.5°N of Moon
18	22:32	LAST QUARTER MOON
21	20:37	Antares 0.8°S of Moon
23	08:04	Saturn 3.3°N of Moon
26	01:32	Partial Solar Eclipse; mag=0.891
26	01:44	NEW MOON
27	17:22	Mercury 3.3°S of Moon
29	19	Mercury at Greatest Elong: 18.4°E
Feb 01	06:34	Moon at Apogee: 404793 km
03	03:09	FIRST QUARTER MOON
04	10:27	Pleiades 0.2°S of Moon
10	14:40	FULL MOON
11	06:15	Regulus 3.8°N of Moon
13	00:46	Moon at Perigee: 363892 km
14	05	Mercury at Inferior Conjunction
14	22:29	Spica 3.4°N of Moon
17	06:42	LAST QUARTER MOON
18	02:00	Antares 0.9°S of Moon
19	18:22	Saturn 3.1°N of Moon
23	04	Uranus at Opposition
23	05:36	Mercury 0.9°N of Moon: Occn.
24	18:26	NEW MOON
25	04	Venus 0.3°S of Mars
Mar 01	00:38	Moon at Apogee: 405695 km
03	18:24	Pleiades 0.1°S of Moon
04	22:52	FIRST QUARTER MOON
08	14	Venus 0.9°N of Jupiter
10	16:57	Regulus 3.8°N of Moon
12	01:37	FULL MOON
12	12	Mercury at Greatest Elong: 27.5°W
13	05:32	Moon at Perigee: 359039 km
14	06:59	Spica 3.2°N of Moon
17	08:08	Antares 1.1°S of Moon
18	16:11	LAST QUARTER MOON
18	17	Mars 0.9°N of Jupiter
19	02:49	Saturn 2.7°N of Moon
20	16:52	Vernal Equinox
26	11:44	NEW MOON
28	11:07	Moon at Apogee: 406463 km
30	01:54	Venus 3.6°S of Moon
31	01:01	Pleiades 0.2°N of Moon

Date	GMT (h:m)	Event
Apr 03	15:11	FIRST QUARTER MOON
07	03:37	Regulus 4.0°N of Moon
09	21:10	Venus 2.4°S of Pleiades
10	10:35	FULL MOON
10	16:08	Moon at Perigee: 356789 km
10	17:39	Spica 3.2°N of Moon
13	16:35	Antares 1.3°S of Moon
15	10:49	Saturn 2.4°N of Moon
17	03:30	LAST QUARTER MOON
22	09	Jupiter in Conjunction with Sun
23	04	Lyrid Meteor Shower
24	13:11	Moon at Apogee: 406676 km
24	14	Mercury at Superior Conjunction
25	04:40	NEW MOON
26	18:12	Mars 4.0°S of Moon
27	06:52	Pleiades 0.3°N of Moon
29	02:39	Venus 0.3°N of Moon: Occn.
May 03	03:26	FIRST QUARTER MOON
04	12:28	Regulus 4.2°N of Moon
05	17	Eta-Aquarid Meteor Shower
08	04:36	Spica 3.2°N of Moon
09	02:43	Moon at Perigee: 357602 km
09	17	Neptune in Conjunction with Sun
09	18:24	FULL MOON
11	02:58	Antares 1.5°S of Moon
12	18:56	Saturn 2.3°N of Moon
16	16:46	LAST QUARTER MOON
21	20:02	Moon at Apogee: 406202 km
23	11	Mercury at Greatest Elong: 22.5°E
24	20:27	NEW MOON
26	19:37	Mercury 0.0°N of Moon: Occn.
28	10	Venus at Greatest Elong: 45.4°E
28	21:26	Venus 3.5°N of Moon
29	04:19	Venus 3.8°S of Pollux
Jun 01	11:54	FIRST QUARTER MOON
04	13:57	Spica 3.3°N of Moon
06	09:14	Moon at Perigee: 361058 km
07	13:32	Antares 1.5°S of Moon
08	02:05	FULL MOON
09	02:38	Saturn 2.3°N of Moon
15	07:45	LAST QUARTER MOON
17	20	Mercury at Inferior Conjunction
18	10:22	Moon at Apogee: 405209 km
20	19:41	Pleiades 0.3°N of Moon
21	10:02	Summer Solstice
22	19	Saturn at Opposition
23	10:36	NEW MOON
23	10:51	Partial Solar Eclipse; mag=0.313
25	03	Mars in Conjunction with Sun
26	22:56	Venus 3.6°N of Moon
30	17:37	FIRST QUARTER MOON

Date	GMT (h:m)	Event
Jul 01	20:58	Spica 3.4°N of Moon
04	04:56	Moon at Perigee: 365988 km
04	22:38	Antares 1.5°S of Moon
05	07	Earth at Aphelion: 1.01667 AU
06	09:06	Saturn 2.5°N of Moon
07	10:33	FULL MOON
07	10:34	Total Lunar Eclipse; mag=1.751
10	23	Mercury at Greatest Elong: 21.1°W
15	00:09	LAST QUARTER MOON
15	07	Saturn at Aphelion: 10.04615 AU
16	03:57	Moon at Apogee: 404370 km
18	03:23	Pleiades 0.3°N of Moon
21	12:31	Mercury 0.9°S of Moon: Occn.
22	22:35	Partial Solar Eclipse; mag=0.360
22	22:49	NEW MOON
24	10:49	Venus 1.8°S of Moon
28	19	Delta-Aquarid Meteor Shower
29	02:24	Spica 3.4°N of Moon
29	22:03	FIRST QUARTER MOON
30	09:01	Moon at Perigee: 369644 km
Aug 01	05:30	Antares 1.5°S of Moon
02	14:10	Saturn 2.5°N of Moon
05	20:38	FULL MOON
06	11	Venus at Inferior Conjunction
07	03	Mercury at Superior Conjunction
12	22:38	Moon at Apogee: 404274 km
13	11	Perseid Meteor Shower
13	17:34	LAST QUARTER MOON
14	11:29	Pleiades 0.4°N of Moon
20	03:25	Mars 3.5°N of Moon
21	09:16	NEW MOON
24	23:21	Moon at Perigee: 366969 km
25	08:12	Spica 3.2°N of Moon
28	02:49	FIRST QUARTER MOON
28	10:55	Antares 1.6°S of Moon
29	18:50	Saturn 2.4°N of Moon
30	08	Uranus in Conjunction with Sun
Sep 04	08:54	FULL MOON
09	17:06	Moon at Apogee: 404975 km
10	19:16	Pleiades 0.6°N of Moon
12	11:18	LAST QUARTER MOON
16	13:37	Venus 2.1°S of Moon
19	18:31	NEW MOON
19	23:56	Mars 0.7°N of Regulus
20	04	Mercury at Greatest Elong: 26.5°E
20	09:41	Mercury 0.6°S of Spica
21	15:28	Moon at Perigee: 361859 km
21	16:04	Spica 3.1°N of Moon
21	17:21	Mercury 2.2°N of Moon
23	02:07	Autumnal Equinox
24	16:48	Antares 1.9°S of Moon
26	01:07	Saturn 2.1°N of Moon
26	09:29	FIRST QUARTER MOON

Date	GMT (h:m)	Event
Oct 03	23:42	FULL MOON
07	08:53	Moon at Apogee: 405929 km
08	02:13	Pleiades 0.8°N of Moon
09	13:35	Venus 1.8°S of Regulus
12	04:22	LAST QUARTER MOON
15	07	Mercury at Inferior Conjunction
15	21:05	Venus 3.8°N of Moon
16	19	Venus at Greatest Elong: 46.3°W
19	03:28	NEW MOON
19	21:58	Moon at Perigee: 357956 km
22	00:56	Antares 2.1°S of Moon
22	03	Orionid Meteor Shower
23	10:54	Saturn 1.6°N of Moon
25	19:13	FIRST QUARTER MOON
30	21	Mercury at Greatest Elong: 18.6°W
Nov 02	16:58	FULL MOON
03	16:20	Moon at Apogee: 406422 km
04	08:27	Pleiades 0.9°N of Moon
06	04	S Taurid Meteor Shower
07	11	Venus 0.1°S of Mars
10	19:39	LAST QUARTER MOON
11	18	Mars at Aphelion: 1.66601 AU
12	16	Neptune at Opposition
13	03	N Taurid Meteor Shower
13	06	Jupiter at Opposition
15	13:33	Spica 3.1°N of Moon
17	09:53	Moon at Perigee: 356788 km
17	12:59	NEW MOON
18	10	Leonid Meteor Shower
20	00:36	Saturn 1.2°N of Moon: Occn.
24	08:41	FIRST QUARTER MOON
30	16:36	Moon at Apogee: 406295 km
Dec 01	14:36	Pleiades 1.0°N of Moon
02	11:55	FULL MOON
06	05	Mercury at Superior Conjunction
10	08:29	LAST QUARTER MOON
12	23:27	Spica 3.2°N of Moon
14	23	Geminid Meteor Shower
15	21:51	Moon at Perigee: 358891 km
16	23:38	NEW MOON
16	23:49	Partial Solar Eclipse; mag=0.882
17	07:58	Mars 3.2°N of Spica
21	23:06	Winter Solstice
23	07	Ursid Meteor Shower
24	01:51	FIRST QUARTER MOON
28	02:49	Moon at Apogee: 405621 km
28	20	Saturn in Conjunction with Sun
28	21:18	Pleiades 0.9°N of Moon

Almanac of Astronomical Events for 2048

Date	GMT (h:m)	Event
Jan 01	06:52	Total Lunar Eclipse; mag=1.128
01	06:57	FULL MOON
03	18	Earth at Perihelion: 0.98328 AU
04	13	Quadrantid Meteor Shower
08	18:49	LAST QUARTER MOON
09	06:36	Spica 3.2°N of Moon
12	08:40	Antares 2.2°S of Moon
13	00:54	Venus 3.0°N of Moon
13	03:53	Moon at Perigee: 363708 km
13	05	Mercury at Greatest Elong: 19.0°E
15	11:32	NEW MOON
16	22:22	Mercury 3.2°S of Moon
22	21:56	FIRST QUARTER MOON
24	21:24	Moon at Apogee: 404717 km
25	04:49	Pleiades 1.0°N of Moon
29	00	Mercury at Inferior Conjunction
30	12	Venus 0.1°S of Saturn
31	00:14	FULL MOON
Feb 05	11:56	Spica 3.1°N of Moon
07	03:16	LAST QUARTER MOON
07	13:42	Mars 4.8°N of Moon
08	15:34	Antares 2.3°S of Moon
09	09:12	Moon at Perigee: 369129 km
10	22:25	Saturn 0.1°N of Moon: Occn.
11	23:13	Venus 1.9°S of Moon
12	05:18	Mercury 0.0°S of Moon: Occn.
14	00:31	NEW MOON
17	15	Mercury 1.4°N of Venus
21	12:51	Pleiades 1.2°N of Moon
21	18:29	Moon at Apogee: 404268 km
21	19:22	FIRST QUARTER MOON
22	22	Mercury at Greatest Elong: 26.6°W
28	04	Uranus at Opposition
29	14:38	FULL MOON
Mar 03	17:58	Spica 2.9°N of Moon
05	02:21	Moon at Perigee: 368850 km
06	16:39	Mars 3.3°N of Moon
06	20:51	Antares 2.5°S of Moon
07	10:45	LAST QUARTER MOON
09	08:40	Saturn 0.3°S of Moon: Occn.
14	14:28	NEW MOON
16	03	Mercury 1.0°S of Venus
19	09:39	Jupiter 4.0°S of Moon
19	20:48	Pleiades 1.4°N of Moon
19	22:34	Vernal Equinox
20	14:41	Moon at Apogee: 404637 km
22	16:03	FIRST QUARTER MOON
30	02:04	FULL MOON
31	02:23	Spica 2.7°N of Moon

Date	GMT (h:m)	Event
Apr 01	10:10	Moon at Perigee: 363563 km
03	03:06	Antares 2.8°S of Moon
03	15:04	Mars 1.7°N of Moon
05	16:30	Saturn 0.7°S of Moon: Occn.
05	18:10	LAST QUARTER MOON
07	14	Mercury at Superior Conjunction
13	05:20	NEW MOON
16	03:47	Jupiter 3.3°S of Moon
16	04:05	Pleiades 1.6°N of Moon
17	07:22	Moon at Apogee: 405534 km
21	10:02	FIRST QUARTER MOON
22	10	Lyrid Meteor Shower
27	12:45	Spica 2.7°N of Moon
28	11:13	FULL MOON
29	14:29	Moon at Perigee: 359129 km
29	15:41	Mercury 1.5°S of Pleiades
30	11:46	Antares 2.9°S of Moon
May 01	03	Mercury 3.2°N of Jupiter
01	05:30	Mars 0.2°N of Moon: Occn.
02	23:38	Saturn 0.9°S of Moon: Occn.
04	09	Mercury at Greatest Elong: 21.0°E
05	00	Eta-Aquarid Meteor Shower
05	02:22	LAST QUARTER MOON
11	07	Neptune in Conjunction with Sun
12	20:58	NEW MOON
14	09:50	Mercury 0.1°N of Moon: Occn.
14	17:19	Moon at Apogee: 406299 km
21	00:16	FIRST QUARTER MOON
24	23:19	Spica 2.8°N of Moon
27	14	Mercury at Inferior Conjunction
27	18:57	FULL MOON
27	22:20	Antares 3.0°S of Moon
27	23:56	Moon at Perigee: 357116 km
28	08:12	Mars 0.9°S of Moon: Occn.
28	20	Venus at Superior Conjunction
29	03	Jupiter in Conjunction with Sun
30	07:08	Saturn 1.0°S of Moon: Occn.
Jun 03	12:05	LAST QUARTER MOON
03	12	Mars at Opposition
09	16:33	Pleiades 1.6°N of Moon
10	20:20	Moon at Apogee: 406480 km
11	12:50	NEW MOON
11	12:57	Annular Solar Eclipse; mag=0.944
15	23:20	Mars 1.7°N of Antares
19	10:49	FIRST QUARTER MOON
20	15:54	Summer Solstice
21	08:21	Spica 2.9°N of Moon
21	20	Mercury at Greatest Elong: 22.7°W
22	15:57	Mercury 2.1°N of Aldebaran
24	05:35	Mars 1.4°S of Moon
24	09:05	Antares 2.9°S of Moon
25	09:51	Moon at Perigee: 357914 km
26	02:01	Partial Lunar Eclipse; mag=0.639
26	02:08	FULL MOON
26	14:52	Saturn 0.9°S of Moon: Occn.
28	02	Mercury 1.9°S of Jupiter

Date	GMT (h:m)	Event
Jul 02	23:58	LAST QUARTER MOON
03	23	Saturn at Opposition
06	05	Earth at Aphelion: 1.01671 AU
06	22:43	Pleiades 1.6°N of Moon
08	03:10	Moon at Apogee: 405968 km
08	12:01	Jupiter 1.4°S of Moon
11	04:04	NEW MOON
18	15:09	Spica 2.8°N of Moon
18	18:31	FIRST QUARTER MOON
21	06	Mercury at Superior Conjunction
21	12:42	Mars 1.7°S of Moon
21	18:16	Antares 3.0°S of Moon
23	16:18	Moon at Perigee: 361240 km
23	22:03	Saturn 0.7°S of Moon: Occn.
25	09:34	FULL MOON
28	01	Delta-Aquarid Meteor Shower
Aug 01	14:30	LAST QUARTER MOON
03	05:34	Pleiades 1.8°N of Moon
03	08:00	Venus 1.0°N of Regulus
04	17:05	Moon at Apogee: 405004 km
05	06:09	Jupiter 0.8°S of Moon: Occn.
05	17:07	Mars 1.1°N of Antares
05	22:46	Mercury 0.8°N of Regulus
09	17:59	NEW MOON
12	13	Mercury 0.8°S of Venus
12	17	Perseid Meteor Shower
14	20:29	Spica 2.6°N of Moon
17	00:32	FIRST QUARTER MOON
18	01:08	Antares 3.2°S of Moon
18	09:21	Mars 2.4°S of Moon
20	04:09	Saturn 0.8°S of Moon: Occn.
20	12:34	Moon at Perigee: 366147 km
23	18:07	FULL MOON
30	13:15	Pleiades 2.0°N of Moon
31	07:42	LAST QUARTER MOON
Sep 01	10:48	Moon at Apogee: 404235 km
01	16	Mercury at Greatest Elong: 27.2°E
01	22:50	Jupiter 0.2°S of Moon: Occn.
03	12	Uranus in Conjunction with Sun
08	06:24	NEW MOON
10	07:03	Mercury 1.3°N of Moon
11	02:13	Spica 2.4°N of Moon
14	06:29	Antares 3.4°S of Moon
15	06:04	FIRST QUARTER MOON
15	15:06	Moon at Perigee: 369916 km
15	15:51	Mars 3.4°S of Moon
16	09:42	Saturn 1.0°S of Moon: Occn.
16	14:16	Venus 2.2°N of Spica
22	04:46	FULL MOON
22	08:01	Autumnal Equinox
26	21:24	Pleiades 2.2°N of Moon
28	00	Mercury at Inferior Conjunction
29	06:20	Moon at Apogee: 404231 km
29	12:30	Jupiter 0.3°N of Moon: Occn.
30	02:45	LAST QUARTER MOON

Date	GMT (h:m)	Event
Oct 03	06	Mars 3.0°S of Saturn
07	17:45	NEW MOON
10	09:41	Venus 1.5°N of Moon
11	04:52	Moon at Perigee: 366937 km
11	12:25	Antares 3.6°S of Moon
13	12	Mercury at Greatest Elong: 18.1°W
13	16:25	Saturn 1.4°S of Moon
14	05:03	Mars 4.8°S of Moon
14	12:20	FIRST QUARTER MOON
20	06	Mars at Perihelion: 1.38142 AU
21	10	Orionid Meteor Shower
21	18:25	FULL MOON
24	05:19	Pleiades 2.4°N of Moon
24	14:31	Venus 2.9°N of Antares
26	21:21	Jupiter 0.6°N of Moon: Occn.
27	01:59	Moon at Apogee: 404996 km
29	22:14	LAST QUARTER MOON
Nov 04	20:01	Spica 2.3°N of Moon
05	10	S Taurid Meteor Shower
06	04:38	NEW MOON
07	20:46	Antares 3.7°S of Moon
07	23:21	Moon at Perigee: 361590 km
09	03:54	Venus 3.1°S of Moon
10	02:19	Saturn 1.8°S of Moon
12	09	N Taurid Meteor Shower
12	20:29	FIRST QUARTER MOON
14	03	Neptune at Opposition
14	12	Mercury at Superior Conjunction
17	16	Leonid Meteor Shower
20	11:20	FULL MOON
20	12:24	Pleiades 2.4°N of Moon
21	11	Venus 2.8°S of Saturn
23	00:36	Jupiter 0.5°N of Moon: Occn.
23	18:24	Moon at Apogee: 405928 km
28	16:34	LAST QUARTER MOON
Dec 02	06:44	Spica 2.4°N of Moon
05	15:30	NEW MOON
05	15:34	Total Solar Eclipse; mag=1.044
06	07:58	Moon at Perigee: 357717 km
07	16:08	Saturn 2.1°S of Moon
12	07:29	FIRST QUARTER MOON
14	05	Geminid Meteor Shower
17	08	Jupiter at Opposition
17	18:37	Pleiades 2.4°N of Moon
19	04	Mercury 2.4°S of Saturn
20	00:01	Jupiter 0.3°N of Moon: Occn.
20	06:26	Pen. Lunar Eclipse; mag=0.962
20	06:39	FULL MOON
21	00:28	Moon at Apogee: 406367 km
21	05:02	Winter Solstice
22	13	Ursid Meteor Shower
26	11	Mercury at Greatest Elong: 19.8°E
28	08:31	LAST QUARTER MOON
29	16:00	Spica 2.4°N of Moon

Almanac of Astronomical Events for 2049

Date	GMT (h:m)	Event
Jan 01	18:56	Antares 3.8°S of Moon
03	00	Venus at Greatest Elong: 47.2°E
03	11	Earth at Perihelion: 0.98334 AU
03	19	Quadrantid Meteor Shower
03	21:03	Moon at Perigee: 356820 km
04	02:24	NEW MOON
08	05	Saturn in Conjunction with Sun
10	21:56	FIRST QUARTER MOON
12	01	Mercury at Inferior Conjunction
14	00:36	Pleiades 2.5°N of Moon
15	23:41	Jupiter 0.1°N of Moon: Occn.
17	00:48	Moon at Apogee: 406228 km
19	02:29	FULL MOON
25	22:42	Spica 2.2°N of Moon
26	21:33	LAST QUARTER MOON
29	04:24	Antares 3.9°S of Moon
31	18:47	Mercury 1.2°S of Moon: Occn.
Feb 01	01:08	Saturn 2.7°S of Moon
01	08:51	Moon at Perigee: 359256 km
02	13:16	NEW MOON
04	07	Mercury at Greatest Elong: 25.4°W
05	07	Mercury 0.4°N of Saturn
05	15:38	Venus 1.8°S of Moon
09	15:38	FIRST QUARTER MOON
10	07:24	Pleiades 2.7°N of Moon
12	03:28	Jupiter 0.2°N of Moon: Occn.
13	12:40	Moon at Apogee: 405515 km
17	20:47	FULL MOON
22	03:59	Spica 1.9°N of Moon
25	07:36	LAST QUARTER MOON
25	11:03	Antares 4.1°S of Moon
28	15:10	Saturn 3.1°S of Moon
Mar 01	12:45	Moon at Perigee: 364178 km
04	00:11	NEW MOON
04	04	Uranus at Opposition
05	03:41	Venus 3.7°N of Moon
07	14:01	Mars 3.2°S of Moon
09	15:31	Pleiades 2.9°N of Moon
11	11:26	FIRST QUARTER MOON
11	13:09	Jupiter 0.6°N of Moon: Occn.
13	07:24	Moon at Apogee: 404598 km
15	22	Venus at Inferior Conjunction
19	12:23	FULL MOON
20	04:28	Vernal Equinox
21	10:04	Spica 1.8°N of Moon
22	05	Mercury at Superior Conjunction
24	16:20	Antares 4.4°S of Moon
26	15:10	LAST QUARTER MOON
28	01:32	Saturn 3.5°S of Moon
28	13:03	Moon at Perigee: 369231 km
31	15:34	Venus 1.5°N of Moon
Apr 02	11:39	NEW MOON
05	11:55	Mars 1.1°S of Moon: Occn.
06	00:19	Pleiades 3.2°N of Moon
08	03:52	Jupiter 1.2°N of Moon
10	03:32	Moon at Apogee: 404203 km
10	07:27	FIRST QUARTER MOON
16	07:53	Mars 3.4°S of Pleiades
16	20	Mercury at Greatest Elong: 19.8°E
17	18:08	Spica 1.7°N of Moon
18	01:04	FULL MOON
22	11:15	Moon at Perigee: 368462 km
22	17	Lyrid Meteor Shower
24	08:54	Saturn 3.8°S of Moon
24	21:11	LAST QUARTER MOON
28	16:23	Venus 4.3°S of Moon
May 02	00:11	NEW MOON
03	08:40	Pleiades 3.2°N of Moon
04	09:50	Mars 1.0°N of Moon: Occn.
05	06	Eta-Aquarid Meteor Shower
05	21:39	Jupiter 1.8°N of Moon
07	14	Mercury at Inferior Conjunction
07	22:34	Moon at Apogee: 404657 km
10	01:57	FIRST QUARTER MOON
13	21	Neptune in Conjunction with Sun
15	03:44	Spica 1.8°N of Moon
17	11:14	FULL MOON
17	11:25	Pen. Lunar Eclipse; mag=0.764
19	19:18	Moon at Perigee: 363447 km
21	15:04	Saturn 3.9°S of Moon
24	02:54	LAST QUARTER MOON
24	23	Venus at Greatest Elong: 45.9°W
31	13:58	Annular Solar Eclipse; mag=0.963
31	14:00	NEW MOON
Jun 02	07:53	Mars 3.0°N of Moon
02	16:42	Jupiter 2.4°N of Moon
03	12	Mercury at Greatest Elong: 24.4°W
04	14:39	Moon at Apogee: 405617 km
08	17:56	FIRST QUARTER MOON
11	13:26	Spica 1.8°N of Moon
11	23	Mars 1.0°N of Jupiter
15	19:13	Pen. Lunar Eclipse; mag=0.251
15	19:26	FULL MOON
16	22:40	Moon at Perigee: 359142 km
17	21:36	Saturn 3.9°S of Moon
18	11:58	Mercury 3.8°N of Aldebaran
20	21:47	Summer Solstice
22	09:41	LAST QUARTER MOON
26	09:59	Venus 4.2°S of Moon
26	21:41	Pleiades 3.2°N of Moon
30	04:50	NEW MOON

63

Date	GMT (h:m)	Event
Jul 02	01:01	Moon at Apogee: 406409 km
04	09	Earth at Aphelion: 1.01667 AU
04	12	Jupiter in Conjunction with Sun
05	15	Mercury at Superior Conjunction
08	07:10	FIRST QUARTER MOON
08	21:48	Spica 1.7°N of Moon
12	02:52	Venus 3.2°N of Aldebaran
15	02:29	FULL MOON
15	04:56	Saturn 3.8°S of Moon
15	07:18	Moon at Perigee: 357060 km
16	04	Saturn at Opposition
21	18:48	LAST QUARTER MOON
24	03:18	Pleiades 3.4°N of Moon
26	08:48	Venus 0.0°S of Moon: Occn.
28	06:59	Jupiter 3.4°N of Moon
28	07	Delta-Aquarid Meteor Shower
29	04:27	Moon at Apogee: 406570 km
29	20:07	NEW MOON
30	00:03	Mercury 0.2°N of Regulus
Aug 04	15	Mars in Conjunction with Sun
05	04:18	Spica 1.5°N of Moon
06	17:52	FIRST QUARTER MOON
11	12:31	Saturn 3.7°S of Moon
12	17:06	Moon at Perigee: 357750 km
12	23	Perseid Meteor Shower
13	09:19	FULL MOON
15	03	Mercury at Greatest Elong: 27.4°E
18	05	Venus 0.6°S of Jupiter
20	07:10	LAST QUARTER MOON
20	09:49	Pleiades 3.6°N of Moon
25	01:31	Jupiter 4.0°N of Moon
25	10:27	Moon at Apogee: 406054 km
25	17:29	Venus 4.3°N of Moon
28	11:18	NEW MOON
30	06:14	Mercury 0.4°N of Moon: Occn.
Sep 01	09:40	Spica 1.2°N of Moon
05	02:28	FIRST QUARTER MOON
07	19:36	Saturn 3.8°S of Moon
08	17	Uranus in Conjunction with Sun
10	00:14	Moon at Perigee: 361122 km
11	07	Mercury at Inferior Conjunction
11	17:04	FULL MOON
16	17:52	Pleiades 3.9°N of Moon
17	19:19	Venus 0.5°N of Regulus
18	23:03	LAST QUARTER MOON
22	00:05	Moon at Apogee: 405133 km
22	13:42	Autumnal Equinox
27	02:05	NEW MOON
27	05	Mercury at Greatest Elong: 17.9°W
27	12	Mercury 0.0°S of Mars
28	15:28	Spica 1.1°N of Moon
28	22	Mars at Aphelion: 1.66617 AU

Date	GMT (h:m)	Event
Oct 04	09:39	FIRST QUARTER MOON
05	02:10	Saturn 4.1°S of Moon
06	07	Venus 0.3°N of Mars
07	21:10	Moon at Perigee: 366276 km
11	02:53	FULL MOON
14	03:07	Pleiades 4.0°N of Moon
18	17:55	LAST QUARTER MOON
19	18:32	Moon at Apogee: 404430 km
21	16	Orionid Meteor Shower
24	12	Neptune at Perihelion: 29.81671 AU
25	16	Mercury at Superior Conjunction
26	16:15	NEW MOON
Nov 01	09:29	Saturn 4.3°S of Moon
02	16:19	FIRST QUARTER MOON
02	18:03	Moon at Perigee: 370118 km
05	16	S Taurid Meteor Shower
09	15:38	FULL MOON
09	15:51	Pen. Lunar Eclipse; mag=0.681
10	12:20	Pleiades 4.1°N of Moon
12	16	N Taurid Meteor Shower
16	15	Neptune at Opposition
16	15:07	Moon at Apogee: 404490 km
17	14:32	LAST QUARTER MOON
17	22	Leonid Meteor Shower
22	08:01	Mars 4.4°N of Moon
22	08:17	Spica 1.1°N of Moon
24	12:14	Mars 2.8°N of Spica
25	05:32	Hybrid Solar Eclipse; mag=1.006
25	05:35	NEW MOON
26	12:29	Mercury 4.3°S of Moon
28	11:08	Moon at Perigee: 366491 km
28	19:26	Saturn 4.5°S of Moon
Dec 01	23:39	FIRST QUARTER MOON
07	20:09	Pleiades 4.1°N of Moon
09	07:28	FULL MOON
09	12	Mercury at Greatest Elong: 20.9°E
14	11	Geminid Meteor Shower
14	11:17	Moon at Apogee: 405260 km
17	11:15	LAST QUARTER MOON
19	17:54	Spica 1.1°N of Moon
21	01:47	Mars 2.6°N of Moon
21	10:51	Winter Solstice
22	19	Ursid Meteor Shower
24	17:51	NEW MOON
26	08:57	Moon at Perigee: 360939 km
26	09:01	Saturn 4.7°S of Moon
27	06	Mercury at Inferior Conjunction
29	21	Venus at Superior Conjunction
31	08:53	FIRST QUARTER MOON

Almanac of Astronomical Events for 2050

Date	GMT (h:m)	Event
Jan 04	01	Quadrantid Meteor Shower
04	02:13	Pleiades 4.1°N of Moon
04	20	Earth at Perihelion: 0.98331 AU
08	01:39	FULL MOON
11	02:36	Moon at Apogee: 406128 km
16	02:12	Spica 0.9°N of Moon
16	06:17	LAST QUARTER MOON
17	15	Mercury at Greatest Elong: 23.9°W
18	18:57	Mars 0.6°N of Moon: Occn.
19	07	Jupiter at Opposition
19	17	Saturn in Conjunction with Sun
21	12:58	Mercury 2.5°S of Moon
23	04:57	NEW MOON
23	18:50	Moon at Perigee: 357295 km
29	20:48	FIRST QUARTER MOON
Feb 06	20:47	FULL MOON
07	06:32	Moon at Apogee: 406519 km
07	22	Mercury 1.1°S of Saturn
12	08:36	Spica 0.6°N of Moon
14	22:10	LAST QUARTER MOON
16	11:04	Mars 1.6°S of Moon
19	17:53	Saturn 5.1°S of Moon
21	07:31	Moon at Perigee: 356855 km
21	15:03	NEW MOON
28	11:29	FIRST QUARTER MOON
28	12:04	Aldebaran 4.1°S of Moon
Mar 05	07	Mercury at Superior Conjunction
06	08:19	Moon at Apogee: 406338 km
08	15:23	FULL MOON
09	05	Uranus at Opposition
11	14:10	Spica 0.4°N of Moon
16	10:08	LAST QUARTER MOON
17	01:42	Mars 3.6°S of Moon
20	10:20	Vernal Equinox
21	17:48	Moon at Perigee: 359625 km
23	00:41	NEW MOON
24	06:48	Mercury 1.4°S of Moon
24	14:14	Venus 3.1°S of Moon
26	21	Mercury 3.7°N of Venus
27	20:22	Aldebaran 3.9°S of Moon
30	04:17	FIRST QUARTER MOON
30	18	Mercury at Greatest Elong: 18.9°E

Date	GMT (h:m)	Event
Apr 02	21:09	Moon at Apogee: 405540 km
07	08:12	FULL MOON
07	20:19	Spica 0.3°N of Moon
14	18:24	LAST QUARTER MOON
18	07	Mercury at Inferior Conjunction
18	19:04	Moon at Perigee: 364528 km
21	10:25	NEW MOON
22	14:45	Venus 3.3°S of Pleiades
22	23	Lyrid Meteor Shower
23	12:53	Venus 1.6°N of Moon
24	05:56	Aldebaran 3.7°S of Moon
28	22:08	FIRST QUARTER MOON
30	15:11	Moon at Apogee: 404600 km
May 05	03:48	Spica 0.4°N of Moon
05	12	Eta-Aquarid Meteor Shower
06	22:26	FULL MOON
06	22:30	Total Lunar Eclipse; mag=1.077
14	00:04	LAST QUARTER MOON
15	15:45	Moon at Perigee: 369166 km
15	18	Mars 1.6°S of Saturn
16	03	Mercury at Greatest Elong: 26.0°W
16	10	Neptune in Conjunction with Sun
20	20:41	Hybrid Solar Eclipse; mag=1.004
20	20:51	NEW MOON
28	10:12	Moon at Apogee: 404241 km
28	16:04	FIRST QUARTER MOON
Jun 01	12:18	Spica 0.3°N of Moon
05	09:51	FULL MOON
09	18:25	Moon at Perigee: 368158 km
12	04:39	LAST QUARTER MOON
15	22	Venus 1.5°N of Jupiter
17	23:07	Aldebaran 3.7°S of Moon
19	08:22	NEW MOON
20	02	Mercury at Superior Conjunction
21	03:33	Summer Solstice
22	12	Neptune at Perihelion: 29.81670 AU
25	04:29	Moon at Apogee: 404736 km
27	09:17	FIRST QUARTER MOON
28	20:52	Spica 0.2°N of Moon

Date	GMT (h:m)	Event
Jul 04	18:51	FULL MOON
06	01	Earth at Aphelion: 1.01663 AU
07	02:25	Moon at Perigee: 363256 km
08	12:50	Venus 0.9°N of Regulus
10	18	Mercury 0.9°N of Jupiter
11	09:46	LAST QUARTER MOON
15	05:16	Aldebaran 3.6°S of Moon
18	21:17	NEW MOON
22	20:27	Moon at Apogee: 405717 km
25	10:44	Mercury 1.0°S of Regulus
26	04:34	Spica 0.0°S of Moon
27	01:05	FIRST QUARTER MOON
28	09	Mercury at Greatest Elong: 27.1°E
28	11	Saturn at Opposition
28	14	Delta-Aquarid Meteor Shower
Aug 03	02:20	FULL MOON
04	05:06	Moon at Perigee: 358975 km
07	19	Jupiter in Conjunction with Sun
07	23	Venus at Greatest Elong: 45.8°E
09	16:48	LAST QUARTER MOON
11	10:39	Aldebaran 3.4°S of Moon
13	06	Perseid Meteor Shower
14	07	Mars at Opposition
17	00	Uranus at Perihelion: 18.28307 AU
17	11:47	NEW MOON
19	07:00	Moon at Apogee: 406488 km
21	14:50	Venus 0.0°N of Moon: Occn.
22	11:01	Spica 0.3°S of Moon
25	02	Mercury at Inferior Conjunction
25	14:56	FIRST QUARTER MOON
Sep 01	09:30	FULL MOON
01	14:03	Moon at Perigee: 356897 km
03	12:12	Venus 1.9°S of Spica
07	07	Mars at Perihelion: 1.38111 AU
07	16:52	Aldebaran 3.2°S of Moon
08	02:51	LAST QUARTER MOON
10	08:36	Mercury 0.3°S of Regulus
10	19	Mercury at Greatest Elong: 18.0°W
13	21	Uranus in Conjunction with Sun
15	09:55	Moon at Apogee: 406591 km
16	03:49	NEW MOON
18	16:45	Spica 0.4°S of Moon
22	19:29	Autumnal Equinox
24	02:34	FIRST QUARTER MOON
30	00:42	Moon at Perigee: 357710 km
30	17:31	FULL MOON

Date	GMT (h:m)	Event
Oct 05	01:05	Aldebaran 3.0°S of Moon
07	00	Mercury at Superior Conjunction
07	16:32	LAST QUARTER MOON
12	15:51	Moon at Apogee: 406050 km
15	20:48	NEW MOON
16	20	Venus at Inferior Conjunction
17	02:49	Jupiter 0.3°N of Regulus
21	22	Orionid Meteor Shower
23	12:10	FIRST QUARTER MOON
28	09:08	Moon at Perigee: 361380 km
30	03:16	FULL MOON
30	03:20	Total Lunar Eclipse; mag=1.054
Nov 01	11:07	Aldebaran 3.0°S of Moon
05	22	S Taurid Meteor Shower
06	09:57	LAST QUARTER MOON
09	06:44	Moon at Apogee: 405118 km
11	15:31	Venus 0.9°N of Moon: Occn.
12	03:23	Mercury 2.1°N of Antares
12	05:48	Spica 0.4°S of Moon
12	22	N Taurid Meteor Shower
14	13:29	Partial Solar Eclipse; mag=0.887
14	13:41	NEW MOON
18	04	Leonid Meteor Shower
19	02	Neptune at Opposition
21	20:25	FIRST QUARTER MOON
22	07	Mercury at Greatest Elong: 22.2°E
25	05:46	Moon at Perigee: 366851 km
27	15:48	Venus 3.3°N of Spica
28	15:10	FULL MOON
28	21:26	Aldebaran 3.0°S of Moon
Dec 06	06:27	LAST QUARTER MOON
07	02:39	Moon at Apogee: 404411 km
09	13:54	Spica 0.5°S of Moon
10	10:27	Venus 3.4°N of Moon
11	13	Mercury at Inferior Conjunction
14	05:18	NEW MOON
14	17	Geminid Meteor Shower
20	16:49	Moon at Perigee: 370280 km
21	04:15	FIRST QUARTER MOON
21	05:31	Mars 4.0°S of Moon
21	16:39	Winter Solstice
23	02	Ursid Meteor Shower
26	06:06	Aldebaran 3.0°S of Moon
27	13	Venus at Greatest Elong: 46.9°W
28	05:15	FULL MOON
31	02	Mercury at Greatest Elong: 22.5°W

Almanac of Astronomical Events for 2051

Date	GMT (h:m)	Event
Jan 03	06	Earth at Perihelion: 0.98330 AU
04	00:04	Moon at Apogee: 404452 km
04	07	Quadrantid Meteor Shower
05	04:29	LAST QUARTER MOON
05	22:16	Spica 0.7°S of Moon
09	02:27	Venus 1.2°N of Moon: Occn.
11	05:42	Mercury 3.6°S of Moon
12	18:58	NEW MOON
15	18:23	Moon at Perigee: 365918 km
18	16:17	Mars 1.6°S of Moon
19	12:38	FIRST QUARTER MOON
22	12:26	Aldebaran 2.8°S of Moon
26	21:20	FULL MOON
31	07	Saturn in Conjunction with Sun
31	19:46	Moon at Apogee: 405173 km
Feb 02	06:01	Spica 1.0°S of Moon
04	01:40	LAST QUARTER MOON
08	01:59	Venus 2.5°S of Moon
11	06:41	NEW MOON
12	18:54	Moon at Perigee: 360557 km
15	16	Mercury at Superior Conjunction
16	04:59	Mars 0.8°N of Moon: Occn.
17	22:16	FIRST QUARTER MOON
18	17:45	Aldebaran 2.6°S of Moon
19	14	Jupiter at Opposition
25	14:53	FULL MOON
28	04:26	Jupiter 0.7°N of Regulus
28	09:20	Moon at Apogee: 405980 km
Mar 01	12:47	Spica 1.2°S of Moon
05	19:47	LAST QUARTER MOON
12	16:52	NEW MOON
13	04	Venus 0.6°N of Saturn
13	05:02	Moon at Perigee: 357407 km
13	23	Mercury at Greatest Elong: 18.3°E
13	23:32	Mercury 0.2°N of Moon: Occn.
14	06	Uranus at Opposition
16	20:00	Mars 2.9°N of Moon
18	00:23	Aldebaran 2.3°S of Moon
19	09:34	FIRST QUARTER MOON
20	15:58	Vernal Equinox
25	10:49	Mars 3.0°S of Pleiades
27	09:00	FULL MOON
27	12:23	Moon at Apogee: 406342 km
28	18:57	Spica 1.3°S of Moon
30	22	Mercury at Inferior Conjunction

Date	GMT (h:m)	Event
Apr 04	09:41	LAST QUARTER MOON
09	23:21	Mercury 2.2°S of Moon
10	16:35	Moon at Perigee: 357337 km
11	01:59	NEW MOON
11	02:09	Partial Solar Eclipse; mag=0.985
14	09:21	Aldebaran 2.2°S of Moon
14	13:12	Mars 4.5°N of Moon
17	22:38	FIRST QUARTER MOON
23	05	Lyrid Meteor Shower
23	15:59	Moon at Apogee: 406100 km
25	01	Mercury 0.9°S of Venus
25	01:06	Spica 1.3°S of Moon
26	02:15	Total Lunar Eclipse; mag=1.202
26	02:19	FULL MOON
27	22	Mercury at Greatest Elong: 27.1°W
May 03	19:30	LAST QUARTER MOON
05	18	Eta-Aquarid Meteor Shower
08	19:01	Mercury 3.8°S of Moon
08	19:23	Venus 2.4°S of Moon
09	01:26	Moon at Perigee: 360140 km
10	10:29	NEW MOON
11	19:46	Aldebaran 2.3°S of Moon
12	05	Mercury 1.1°S of Venus
17	13:29	FIRST QUARTER MOON
19	00	Neptune in Conjunction with Sun
21	04:56	Moon at Apogee: 405251 km
22	07:47	Spica 1.3°S of Moon
25	17:35	FULL MOON
Jun 02	02:15	LAST QUARTER MOON
04	14	Mercury at Superior Conjunction
06	01:20	Moon at Perigee: 364806 km
07	13:52	Venus 1.6°N of Moon
08	18:56	NEW MOON
14	00:25	Jupiter 0.5°N of Regulus
16	05:55	FIRST QUARTER MOON
17	22:14	Moon at Apogee: 404346 km
18	15:08	Spica 1.5°S of Moon
21	09:17	Summer Solstice
24	06:14	FULL MOON
29	19	Mercury 0.2°N of Mars

Date	GMT (h:m)	Event
Jul 01	07:15	LAST QUARTER MOON
02	20:59	Moon at Perigee: 369221 km
05	14:03	Aldebaran 2.2°S of Moon
06	10	Earth at Aphelion: 1.01673 AU
08	04:09	NEW MOON
10	09	Mercury at Greatest Elong: 26.3°E
15	16:49	Moon at Apogee: 404065 km
15	22:51	Spica 1.7°S of Moon
15	23:21	FIRST QUARTER MOON
23	16:36	FULL MOON
28	00:46	Moon at Perigee: 368220 km
28	20	Delta-Aquarid Meteor Shower
30	11:52	LAST QUARTER MOON
Aug 01	20:15	Aldebaran 2.0°S of Moon
05	07	Venus at Superior Conjunction
06	15:05	NEW MOON
07	06	Mercury at Inferior Conjunction
09	21	Saturn at Opposition
12	06:29	Spica 2.0°S of Moon
12	11:22	Moon at Apogee: 404639 km
13	12	Perseid Meteor Shower
14	16:49	FIRST QUARTER MOON
16	21	Mars at Aphelion: 1.66620 AU
22	01:35	FULL MOON
24	08:43	Moon at Perigee: 363298 km
25	05	Mercury at Greatest Elong: 18.4°W
28	17:29	LAST QUARTER MOON
29	01:34	Aldebaran 1.8°S of Moon
Sep 05	04:33	NEW MOON
08	13:34	Spica 2.1°S of Moon
09	02	Jupiter in Conjunction with Sun
09	04:18	Moon at Apogee: 405670 km
10	12	Mars in Conjunction with Sun
13	09:20	FIRST QUARTER MOON
19	00	Uranus in Conjunction with Sun
19	08	Mercury at Superior Conjunction
20	10:11	FULL MOON
21	12:20	Moon at Perigee: 358920 km
23	01:26	Autumnal Equinox
25	07:58	Aldebaran 1.7°S of Moon
27	01:22	LAST QUARTER MOON

Date	GMT (h:m)	Event
Oct 01	15:34	Venus 2.6°N of Spica
03	07:52	Jupiter 4.0°N of Moon
04	20:47	NEW MOON
04	21:01	Partial Solar Eclipse; mag=0.602
06	09:52	Venus 0.0°N of Moon: Occn.
06	15:34	Moon at Apogee: 406415 km
13	00:12	FIRST QUARTER MOON
19	19:10	Total Lunar Eclipse; mag=1.412
19	19:13	FULL MOON
19	22:41	Moon at Perigee: 356809 km
20	21	Mercury 2.0°S of Venus
22	04	Orionid Meteor Shower
22	16:49	Aldebaran 1.6°S of Moon
26	12:39	LAST QUARTER MOON
31	00:39	Jupiter 3.6°N of Moon
Nov 02	00:12	Mars 0.9°N of Moon: Occn.
02	02:08	Spica 2.2°S of Moon
02	17:22	Moon at Apogee: 406481 km
03	14:59	NEW MOON
04	21	Mercury at Greatest Elong: 23.5°E
05	05:54	Mars 2.6°N of Spica
05	18:45	Venus 4.3°S of Moon
06	05	S Taurid Meteor Shower
07	17:11	Venus 3.8°N of Antares
09	09:41	Mercury 1.8°N of Antares
11	13:07	FIRST QUARTER MOON
13	04	N Taurid Meteor Shower
17	10:55	Moon at Perigee: 357776 km
18	05:06	FULL MOON
18	10	Leonid Meteor Shower
19	03:44	Aldebaran 1.7°S of Moon
21	14	Neptune at Opposition
25	04:02	LAST QUARTER MOON
25	20	Mercury at Inferior Conjunction
27	16:03	Jupiter 3.2°N of Moon
29	08:27	Spica 2.2°S of Moon
29	23:42	Moon at Apogee: 405942 km
30	20:01	Mars 1.1°S of Moon: Occn.
Dec 03	09:37	NEW MOON
11	00:07	FIRST QUARTER MOON
13	18	Mercury at Greatest Elong: 21.1°W
14	23	Geminid Meteor Shower
15	19:57	Moon at Perigee: 361745 km
16	14:46	Aldebaran 1.7°S of Moon
17	16:05	FULL MOON
21	22:33	Winter Solstice
23	08	Ursid Meteor Shower
24	23:21	LAST QUARTER MOON
25	05:28	Jupiter 2.8°N of Moon
26	15:27	Spica 2.4°S of Moon
27	16:05	Moon at Apogee: 405014 km
29	17:58	Mars 3.0°S of Moon
31	18:31	Mercury 4.5°S of Moon

Almanac of Astronomical Events for 2052

Date	GMT (h:m)	Event
Jan 02	03	Venus 0.8°S of Saturn
02	03:05	NEW MOON
04	14	Quadrantid Meteor Shower
05	10	Earth at Perihelion: 0.98334 AU
09	09:27	FIRST QUARTER MOON
12	14:41	Moon at Perigee: 367399 km
12	23:41	Aldebaran 1.6°S of Moon
16	04:24	FULL MOON
21	15:54	Jupiter 2.5°N of Moon
22	23:18	Spica 2.7°S of Moon
23	21:03	LAST QUARTER MOON
24	12:51	Moon at Apogee: 404327 km
27	18:09	Mars 4.6°S of Moon
28	04	Mercury at Superior Conjunction
31	18:30	NEW MOON
Feb 04	04:10	Venus 2.2°S of Moon
06	18:01	Moon at Perigee: 370138 km
07	17:35	FIRST QUARTER MOON
09	05:56	Aldebaran 1.3°S of Moon
12	03	Saturn in Conjunction with Sun
14	18:21	FULL MOON
17	22:13	Jupiter 2.5°N of Moon
19	07:33	Spica 3.0°S of Moon
21	10:01	Moon at Apogee: 404402 km
22	18:44	LAST QUARTER MOON
25	10	Mercury at Greatest Elong: 18.1°E
Mar 01	07:36	NEW MOON
02	10:24	Mercury 1.4°N of Moon
04	04:33	Moon at Perigee: 365498 km
04	15:28	Venus 3.4°N of Moon
07	11:16	Aldebaran 1.2°S of Moon
08	01:17	FIRST QUARTER MOON
12	08	Mercury at Inferior Conjunction
15	09:54	FULL MOON
15	11	Venus at Greatest Elong: 46.2°E
16	00:29	Jupiter 2.8°N of Moon
17	15:21	Spica 3.1°S of Moon
18	07	Uranus at Opposition
19	21:56	Vernal Equinox
20	04:36	Moon at Apogee: 405152 km
21	02	Jupiter at Opposition
23	14:09	LAST QUARTER MOON
29	01:25	Mercury 2.8°S of Moon
30	18:27	NEW MOON
30	18:30	Total Solar Eclipse; mag=1.047

Date	GMT (h:m)	Event
Apr 01	05:25	Moon at Perigee: 360446 km
03	18:08	Aldebaran 1.1°S of Moon
04	15:10	Venus 0.3°N of Pleiades
06	09:28	FIRST QUARTER MOON
09	00	Mercury at Greatest Elong: 27.8°W
12	00:43	Jupiter 3.1°N of Moon
13	22:11	Spica 3.0°S of Moon
14	02:16	Pen. Lunar Eclipse; mag=0.947
14	02:29	FULL MOON
16	17:03	Moon at Apogee: 405976 km
22	06:03	LAST QUARTER MOON
22	11	Lyrid Meteor Shower
27	17:17	Mercury 2.3°S of Moon
29	03:20	NEW MOON
29	14:35	Moon at Perigee: 357547 km
May 01	03:27	Aldebaran 1.1°S of Moon
05	00	Eta-Aquarid Meteor Shower
05	19:05	FIRST QUARTER MOON
07	00	Mars 0.5°S of Saturn
09	02:21	Jupiter 3.2°N of Moon
11	04:10	Spica 3.1°S of Moon
13	19:00	FULL MOON
13	20:41	Moon at Apogee: 406342 km
19	01	Mercury at Superior Conjunction
20	14	Neptune in Conjunction with Sun
21	18:15	LAST QUARTER MOON
25	11	Venus at Inferior Conjunction
28	00:39	Moon at Perigee: 357494 km
28	10:50	NEW MOON
Jun 03	06:21	Regulus 4.1°N of Moon
04	06:49	FIRST QUARTER MOON
05	08:20	Jupiter 2.9°N of Moon
07	10:04	Spica 3.2°S of Moon
10	00:46	Moon at Apogee: 406066 km
12	10:27	FULL MOON
20	03:10	LAST QUARTER MOON
20	09:46	Mars 3.5°S of Moon
20	15:16	Summer Solstice
21	02	Mercury at Greatest Elong: 25.0°E
24	05:13	Venus 0.2°N of Moon: Occn.
25	00:31	Aldebaran 1.1°S of Moon
25	08:22	Moon at Perigee: 360167 km
26	17:50	NEW MOON
28	12:01	Mercury 3.9°N of Moon
30	14:54	Regulus 3.8°N of Moon

Date	GMT (h:m)	Event
Jul 02	19:31	Jupiter 2.3°N of Moon
03	17	Earth at Aphelion: 1.01670 AU
03	20:59	FIRST QUARTER MOON
04	16:42	Spica 3.4°S of Moon
07	12:41	Moon at Apogee: 405232 km
12	00:23	FULL MOON
15	08:30	Saturn 4.7°S of Moon
15	12:01	Venus 1.3°N of Aldebaran
18	14	Mercury at Inferior Conjunction
18	21:48	Mars 2.1°S of Moon
19	09:37	LAST QUARTER MOON
22	08:57	Aldebaran 1.0°S of Moon
22	18:08	Venus 0.7°N of Moon: Occn.
23	08:05	Moon at Perigee: 364746 km
25	13	Mars at Perihelion: 1.38118 AU
26	01:31	NEW MOON
28	00:25	Regulus 3.7°N of Moon
28	02	Delta-Aquarid Meteor Shower
30	10:57	Jupiter 1.7°N of Moon
Aug 01	00:26	Spica 3.7°S of Moon
02	13:20	FIRST QUARTER MOON
03	17	Venus at Greatest Elong: 45.8°W
04	05:18	Moon at Apogee: 404407 km
07	06	Mercury at Greatest Elong: 19.2°W
10	12:52	FULL MOON
12	18	Perseid Meteor Shower
16	03:24	Mars 0.9°S of Moon: Occn.
17	14:43	LAST QUARTER MOON
18	15:12	Aldebaran 0.8°S of Moon
19	04:12	Moon at Perigee: 369240 km
21	00:48	Venus 2.8°N of Moon
21	10	Saturn at Opposition
24	11:07	NEW MOON
27	04:55	Jupiter 1.0°N of Moon: Occn.
28	08:51	Spica 3.9°S of Moon
Sep 01	00:01	Moon at Apogee: 404226 km
01	07:10	FIRST QUARTER MOON
01	10	Mercury at Superior Conjunction
09	00:15	FULL MOON
12	22:54	Mars 0.0°S of Moon: Occn.
13	06:22	Moon at Perigee: 368182 km
14	20:33	Aldebaran 0.7°S of Moon
15	19:48	LAST QUARTER MOON
19	17:34	Venus 3.8°N of Moon
20	17:10	Regulus 3.7°N of Moon
22	07:16	Autumnal Equinox
22	23:32	NEW MOON
22	23:38	Annular Solar Eclipse; mag=0.973
23	04	Uranus in Conjunction with Sun
24	09:51	Mercury 1.8°S of Moon
24	17:05	Spica 3.9°S of Moon
27	11:11	Mercury 1.1°N of Spica
28	19:24	Moon at Apogee: 404888 km
30	22:28	Venus 0.0°S of Regulus

Date	GMT (h:m)	Event
Oct 01	01:36	FIRST QUARTER MOON
07	22	Jupiter at Aphelion: 5.45334 AU
08	10:44	Partial Lunar Eclipse; mag=0.082
08	10:54	FULL MOON
09	00	Jupiter in Conjunction with Sun
10	04:04	Mars 0.6°N of Moon: Occn.
10	15:30	Moon at Perigee: 362981 km
12	03:06	Aldebaran 0.7°S of Moon
15	02:22	LAST QUARTER MOON
17	10	Mercury at Greatest Elong: 24.8°E
17	23:05	Regulus 3.7°N of Moon
19	16:13	Venus 2.6°N of Moon
21	10	Orionid Meteor Shower
22	15:03	NEW MOON
26	13:06	Moon at Apogee: 405938 km
28	07	Mars at Opposition
30	19:39	FIRST QUARTER MOON
Nov 05	11	S Taurid Meteor Shower
06	00:24	Mars 1.5°N of Moon
06	21:09	FULL MOON
07	21:02	Moon at Perigee: 358409 km
08	12:17	Aldebaran 0.8°S of Moon
09	02	Mercury at Inferior Conjunction
12	10	N Taurid Meteor Shower
12	11:22	Jupiter 3.0°N of Spica
13	11:50	LAST QUARTER MOON
14	04:31	Regulus 3.5°N of Moon
14	12:52	Venus 3.5°N of Spica
14	22	Venus 0.6°N of Jupiter
17	16	Leonid Meteor Shower
18	06:19	Spica 3.9°S of Moon
18	11:01	Jupiter 0.7°S of Moon: Occn.
18	19:20	Venus 0.5°S of Moon: Occn.
19	17:31	Mercury 0.7°S of Moon: Occn.
21	09:02	NEW MOON
22	23:40	Moon at Apogee: 406619 km
23	01	Neptune at Opposition
25	18	Mercury at Greatest Elong: 19.9°W
29	12:16	FIRST QUARTER MOON
Dec 03	04:58	Mars 2.8°N of Moon
05	23:33	Aldebaran 0.8°S of Moon
06	07:18	FULL MOON
06	08:52	Moon at Perigee: 356425 km
11	11:26	Regulus 3.3°N of Moon
13	01:07	LAST QUARTER MOON
14	05	Geminid Meteor Shower
16	02:31	Jupiter 1.3°S of Moon
19	04:33	Venus 3.6°S of Moon
20	00:02	Moon at Apogee: 406649 km
21	04:15	NEW MOON
21	04:18	Winter Solstice
22	14	Ursid Meteor Shower
26	02:34	Saturn 4.3°S of Moon
29	02:28	FIRST QUARTER MOON
30	22:54	Mars 4.3°N of Moon

Almanac of Astronomical Events for 2053

Date	GMT (h:m)	Event
Jan 02	10:41	Aldebaran 0.7°S of Moon
03	20	Quadrantid Meteor Shower
03	21:39	Moon at Perigee: 357808 km
03	22	Earth at Perihelion: 0.98330 AU
04	17:46	FULL MOON
07	13	Mercury at Superior Conjunction
07	20:49	Regulus 3.0°N of Moon
11	18:09	LAST QUARTER MOON
12	16:25	Jupiter 1.8°S of Moon
16	07:55	Moon at Apogee: 406076 km
19	23:12	NEW MOON
22	13:19	Saturn 3.8°S of Moon
27	13:41	FIRST QUARTER MOON
29	19:30	Aldebaran 0.6°S of Moon
Feb 01	05:49	Moon at Perigee: 362186 km
03	04:57	FULL MOON
03	08	Mercury 1.0°N of Saturn
04	07:34	Regulus 2.9°N of Moon
07	23	Mercury at Greatest Elong: 18.2°E
09	04:09	Jupiter 2.0°S of Moon
10	13:49	LAST QUARTER MOON
13	01:22	Moon at Apogee: 405090 km
18	16:31	NEW MOON
23	04	Saturn in Conjunction with Sun
23	10	Mercury at Inferior Conjunction
23	15:30	Mars 2.4°S of Pleiades
25	22:09	FIRST QUARTER MOON
26	01:40	Aldebaran 0.4°S of Moon
28	20:20	Moon at Perigee: 367800 km
Mar 03	17:31	Regulus 2.9°N of Moon
04	17:09	FULL MOON
04	17:20	Pen. Lunar Eclipse; mag=0.932
08	12:38	Jupiter 2.0°S of Moon
12	10:21	LAST QUARTER MOON
12	21:51	Moon at Apogee: 404382 km
16	11	Venus at Superior Conjunction
18	03:05	Mercury 2.8°S of Moon
18	16:16	Saturn 3.3°S of Moon
20	03:46	Vernal Equinox
20	07:07	Annular Solar Eclipse; mag=0.992
20	07:11	NEW MOON
22	07	Mercury at Greatest Elong: 27.7°W
23	08	Uranus at Opposition
25	07:05	Aldebaran 0.4°S of Moon
25	21:13	Moon at Perigee: 369681 km
25	22	Mercury 0.1°S of Saturn
27	04:50	FIRST QUARTER MOON
31	01:09	Regulus 2.9°N of Moon

Date	GMT (h:m)	Event
Apr 03	06:22	FULL MOON
04	17:05	Jupiter 1.8°S of Moon
09	17:46	Moon at Apogee: 404471 km
11	06:04	LAST QUARTER MOON
15	06:55	Saturn 3.1°S of Moon
17	13:52	Mercury 0.8°S of Moon: Occn.
18	18:48	NEW MOON
20	17	Jupiter at Opposition
21	13:57	Moon at Perigee: 365050 km
21	14:01	Aldebaran 0.5°S of Moon
22	17	Lyrid Meteor Shower
25	11:02	FIRST QUARTER MOON
27	06:47	Regulus 2.8°N of Moon
May 01	18:15	Jupiter 1.5°S of Moon
02	20:25	FULL MOON
03	09	Mercury at Superior Conjunction
05	06	Eta-Aquarid Meteor Shower
07	10:56	Moon at Apogee: 405235 km
10	23:40	LAST QUARTER MOON
12	20:06	Saturn 2.8°S of Moon
17	17	Mercury 1.5°N of Venus
18	03:43	NEW MOON
19	14:01	Moon at Perigee: 360272 km
23	04	Neptune in Conjunction with Sun
24	12:16	Regulus 2.6°N of Moon
24	18:04	FIRST QUARTER MOON
28	18:58	Jupiter 1.5°S of Moon

Date	GMT (h:m)	Event
Jun 01	11:02	FULL MOON
02	16	Mercury at Greatest Elong: 23.4°E
03	22:38	Moon at Apogee: 406060 km
09	00	Mercury 0.8°S of Venus
09	06:11	Saturn 2.5°S of Moon
09	14:19	LAST QUARTER MOON
15	09:46	Aldebaran 0.5°S of Moon
16	10:51	NEW MOON
16	21:48	Moon at Perigee: 357537 km
17	13:49	Mercury 3.1°N of Moon
19	07:38	Mars 4.8°N of Moon
20	19:22	Regulus 2.3°N of Moon
20	21:03	Summer Solstice
23	02:55	FIRST QUARTER MOON
24	22:39	Jupiter 1.7°S of Moon
29	01	Mercury at Inferior Conjunction

Date	GMT (h:m)	Event
Jul 01	02:01	FULL MOON
01	02:40	Moon at Apogee: 406396 km
03	22	Mars at Aphelion: 1.66610 AU
06	12:35	Saturn 2.4°S of Moon
06	13	Earth at Aphelion: 1.01667 AU
09	01:47	LAST QUARTER MOON
12	19:56	Aldebaran 0.4°S of Moon
14	11:00	Mercury 1.1°N of Moon: Occn.
14	12	Venus 0.5°N of Mars
15	07:11	Moon at Perigee: 357527 km
15	17:26	NEW MOON
17	22:05	Mars 3.2°N of Moon
18	01:50	Venus 3.5°N of Moon
18	04:36	Regulus 2.1°N of Moon
19	21:07	Venus 1.0°N of Regulus
20	21	Mercury at Greatest Elong: 20.3°W
22	07:25	Jupiter 2.2°S of Moon
22	14:16	FIRST QUARTER MOON
24	19:46	Mars 0.6°N of Regulus
28	06:44	Moon at Apogee: 406073 km
28	08	Delta-Aquarid Meteor Shower
30	17:06	FULL MOON
Aug 02	16:08	Saturn 2.4°S of Moon
07	10:25	LAST QUARTER MOON
09	04:19	Aldebaran 0.3°S of Moon
12	14:58	Moon at Perigee: 360223 km
13	00	Perseid Meteor Shower
14	00:41	NEW MOON
15	13:13	Mars 1.3°N of Moon
16	01	Mercury at Superior Conjunction
16	20:47	Venus 1.0°S of Moon: Occn.
18	21:20	Jupiter 2.7°S of Moon
21	04:27	FIRST QUARTER MOON
24	18:27	Moon at Apogee: 405238 km
29	07:53	FULL MOON
29	08:04	Pen. Lunar Eclipse; mag=1.019
29	18:48	Saturn 2.6°S of Moon
Sep 03	05	Saturn at Opposition
03	23:20	Venus 1.3°N of Spica
05	10:36	Aldebaran 0.2°S of Moon
05	17:05	LAST QUARTER MOON
09	15:22	Moon at Perigee: 364934 km
11	00:57	Regulus 2.1°N of Moon
12	09:32	Total Solar Eclipse; mag=1.033
12	09:36	NEW MOON
13	11	Venus 2.3°S of Jupiter
13	23:29	Mercury 3.4°S of Moon
15	15:00	Jupiter 3.1°S of Moon
19	21:29	FIRST QUARTER MOON
21	11:30	Moon at Apogee: 404434 km
21	17:27	Mercury 0.2°N of Spica
22	13:05	Autumnal Equinox
25	22:39	Saturn 2.8°S of Moon
27	21:50	FULL MOON
28	07	Uranus in Conjunction with Sun
29	22	Mercury at Greatest Elong: 26.0°E

Date	GMT (h:m)	Event
Oct 02	16:00	Aldebaran 0.3°S of Moon
04	23:01	LAST QUARTER MOON
06	10:09	Moon at Perigee: 369576 km
08	08:54	Regulus 2.0°N of Moon
11	20:53	NEW MOON
13	10:29	Jupiter 3.4°S of Moon
13	13	Mercury 4.3°S of Jupiter
15	23:07	Venus 1.1°N of Antares
16	17	Mars in Conjunction with Sun
19	07:14	Moon at Apogee: 404289 km
19	16:54	FIRST QUARTER MOON
20	05	Venus at Greatest Elong: 46.9°E
21	16	Orionid Meteor Shower
23	04:45	Saturn 2.8°S of Moon
24	05	Mercury at Inferior Conjunction
27	10:38	FULL MOON
29	22:36	Aldebaran 0.4°S of Moon
31	10:02	Moon at Perigee: 368005 km
Nov 03	05:37	LAST QUARTER MOON
04	14:40	Regulus 1.9°N of Moon
05	17	S Taurid Meteor Shower
08	09	Jupiter in Conjunction with Sun
08	20:11	Mercury 1.3°S of Moon
09	02	Mercury at Greatest Elong: 19.0°W
10	10:55	NEW MOON
12	16	N Taurid Meteor Shower
16	03:37	Moon at Apogee: 404957 km
17	22	Leonid Meteor Shower
18	13:26	FIRST QUARTER MOON
19	12:59	Saturn 2.6°S of Moon
25	12	Neptune at Opposition
25	22:21	FULL MOON
26	07:44	Aldebaran 0.5°S of Moon
27	23:02	Moon at Perigee: 362464 km
Dec 01	20:00	Regulus 1.6°N of Moon
02	14:04	LAST QUARTER MOON
08	00:12	Jupiter 3.9°S of Moon
10	03:40	NEW MOON
13	21:20	Moon at Apogee: 405943 km
14	12	Geminid Meteor Shower
16	22:37	Saturn 2.1°S of Moon
17	22	Mercury at Superior Conjunction
18	09:11	FIRST QUARTER MOON
21	10:09	Winter Solstice
22	20	Ursid Meteor Shower
23	18:42	Aldebaran 0.4°S of Moon
25	09:23	FULL MOON
26	06:40	Moon at Perigee: 358032 km
29	03:21	Regulus 1.4°N of Moon
30	02	Venus at Inferior Conjunction

Almanac of Astronomical Events for 2054

Date	GMT (h:m)	Event
Jan 01	01:11	LAST QUARTER MOON
02	17	Earth at Perihelion: 0.98333 AU
04	02	Quadrantid Meteor Shower
04	16:39	Jupiter 4.2°S of Moon
08	22:34	NEW MOON
10	05:56	Moon at Apogee: 406537 km
13	09:11	Saturn 1.6°S of Moon
17	02:14	FIRST QUARTER MOON
20	05:22	Aldebaran 0.3°S of Moon
22	10	Mercury at Greatest Elong: 18.6°E
23	19:38	Moon at Perigee: 356512 km
23	20:08	FULL MOON
25	13:33	Regulus 1.2°N of Moon
30	15:08	LAST QUARTER MOON
Feb 01	07:20	Jupiter 4.4°S of Moon
04	01:44	Venus 1.1°N of Moon: Occn.
06	06:08	Moon at Apogee: 406527 km
06	23	Mercury at Inferior Conjunction
07	18:14	NEW MOON
09	20:43	Saturn 1.3°S of Moon
15	15:36	FIRST QUARTER MOON
16	13:48	Aldebaran 0.2°S of Moon
21	07:51	Moon at Perigee: 358350 km
22	01:02	Regulus 1.2°N of Moon
22	06:46	FULL MOON
22	06:50	Total Lunar Eclipse; mag=1.277
28	20:08	Jupiter 4.4°S of Moon
Mar 01	07:36	LAST QUARTER MOON
04	17	Mercury at Greatest Elong: 27.2°W
05	06:37	Venus 0.3°S of Moon: Occn.
05	16:18	Moon at Apogee: 405875 km
05	18:58	Mars 4.0°S of Moon
07	03:26	Mercury 2.4°S of Moon
07	14	Saturn in Conjunction with Sun
09	12:32	Partial Solar Eclipse; mag=0.668
09	12:46	NEW MOON
10	15	Venus at Greatest Elong: 46.7°W
15	19:53	Aldebaran 0.2°S of Moon
17	01:21	FIRST QUARTER MOON
20	09:35	Vernal Equinox
21	11:17	Regulus 1.2°N of Moon
21	14:11	Moon at Perigee: 362830 km
23	17:21	FULL MOON
25	18	Venus 2.3°N of Mars
28	05	Mercury 0.5°S of Saturn
28	06:16	Jupiter 4.3°S of Moon
28	10	Uranus at Opposition
31	01:50	LAST QUARTER MOON

Date	GMT (h:m)	Event
Apr 02	10:08	Moon at Apogee: 404828 km
03	22:54	Mars 2.4°S of Moon
04	05:26	Venus 0.5°S of Moon: Occn.
05	22:54	Saturn 0.7°S of Moon: Occn.
08	04:32	NEW MOON
12	01:22	Aldebaran 0.3°S of Moon
15	08:23	FIRST QUARTER MOON
17	12	Mercury at Superior Conjunction
17	18:52	Regulus 1.1°N of Moon
18	00:58	Moon at Perigee: 368088 km
22	04:02	FULL MOON
22	23	Lyrid Meteor Shower
24	11	Venus 0.8°N of Saturn
24	12:42	Jupiter 4.1°S of Moon
29	20:46	LAST QUARTER MOON
30	05:45	Moon at Apogee: 404138 km
May 03	02:04	Mars 0.4°S of Moon: Occn.
03	10:37	Mercury 2.0°S of Pleiades
03	12:30	Saturn 0.3°S of Moon: Occn.
04	07:58	Venus 0.9°N of Moon: Occn.
05	13	Eta-Aquarid Meteor Shower
07	17:00	NEW MOON
09	08:10	Aldebaran 0.5°S of Moon
10	23	Mars 0.4°N of Saturn
13	03:26	Moon at Perigee: 369431 km
14	13:57	FIRST QUARTER MOON
15	00:22	Regulus 0.9°N of Moon
15	10	Mercury at Greatest Elong: 21.8°E
21	15:16	FULL MOON
21	15:23	Jupiter 4.1°S of Moon
22	08	Jupiter at Opposition
25	18	Neptune in Conjunction with Sun
28	00:43	Moon at Apogee: 404285 km
29	15:03	LAST QUARTER MOON
31	00:51	Saturn 0.0°S of Moon: Occn.
Jun 01	03:57	Mars 1.5°N of Moon
03	09:44	Venus 2.8°N of Moon
06	02:40	NEW MOON
08	21:43	Moon at Perigee: 365007 km
08	22	Mercury at Inferior Conjunction
11	05:50	Regulus 0.6°N of Moon
12	11	Mars at Perihelion: 1.38142 AU
12	19:17	FIRST QUARTER MOON
17	16:10	Jupiter 4.2°S of Moon
20	03:42	FULL MOON
21	02:47	Summer Solstice
24	17:32	Moon at Apogee: 405098 km
27	10:39	Saturn 0.2°N of Moon: Occn.
28	07:30	LAST QUARTER MOON
30	04:02	Mars 3.1°N of Moon

Date	GMT (h:m)	Event
Jul 03	00	Mercury at Greatest Elong: 21.8°W
03	02:49	Aldebaran 0.4°S of Moon
03	10:39	Venus 3.9°N of Moon
03	20:42	Mercury 1.8°N of Moon
05	10:34	NEW MOON
05	16	Earth at Aphelion: 1.01672 AU
06	21:10	Moon at Perigee: 360411 km
08	13:10	Regulus 0.4°N of Moon
12	01:36	FIRST QUARTER MOON
14	18:17	Jupiter 4.5°S of Moon
19	17:47	FULL MOON
22	05:49	Moon at Apogee: 405947 km
24	17:10	Saturn 0.2°N of Moon: Occn.
27	21:27	LAST QUARTER MOON
28	14	Delta-Aquarid Meteor Shower
29	01:20	Mars 3.9°N of Moon
30	12:28	Aldebaran 0.3°S of Moon
31	01	Mercury at Superior Conjunction
Aug 02	10:10	Venus 3.4°N of Moon
03	17:48	NEW MOON
03	18:02	Partial Solar Eclipse; mag=0.066
04	04:47	Moon at Perigee: 357668 km
04	22:48	Regulus 0.4°N of Moon
10	10:05	FIRST QUARTER MOON
11	00:42	Jupiter 4.7°S of Moon
13	06	Perseid Meteor Shower
18	09:22	FULL MOON
18	09:25	Total Lunar Eclipse; mag=1.306
18	10:18	Moon at Apogee: 406258 km
20	20:46	Saturn 0.0°N of Moon: Occn.
26	08:56	LAST QUARTER MOON
26	18:13	Mars 4.2°N of Moon
26	20:35	Aldebaran 0.3°S of Moon
Sep 01	14:49	Moon at Perigee: 357587 km
02	01:08	Partial Solar Eclipse; mag=0.979
02	01:18	NEW MOON
03	17:20	Mercury 4.9°S of Moon
07	12:38	Jupiter 4.7°S of Moon
08	21:46	FIRST QUARTER MOON
12	10	Mercury at Greatest Elong: 26.9°E
14	13:46	Moon at Apogee: 405922 km
16	04	Saturn at Opposition
16	23:01	Saturn 0.2°S of Moon: Occn.
17	01:41	FULL MOON
22	17:34	Mercury 1.4°S of Spica
22	19:00	Autumnal Equinox
23	02:49	Aldebaran 0.4°S of Moon
24	04:11	Mars 4.2°N of Moon
24	18:26	LAST QUARTER MOON
28	19:54	Regulus 0.3°N of Moon
29	23:51	Moon at Perigee: 360310 km

Date	GMT (h:m)	Event
Oct 01	09:49	NEW MOON
03	09	Uranus in Conjunction with Sun
05	05:25	Jupiter 4.7°S of Moon
08	02	Mercury at Inferior Conjunction
08	13:19	FIRST QUARTER MOON
12	01:43	Moon at Apogee: 405118 km
13	06	Venus at Superior Conjunction
14	01:59	Saturn 0.3°S of Moon: Occn.
16	17:44	FULL MOON
20	08:19	Aldebaran 0.5°S of Moon
21	23	Orionid Meteor Shower
22	03:36	Mars 4.4°N of Moon
23	15	Mercury at Greatest Elong: 18.3°W
24	02:39	LAST QUARTER MOON
26	03:50	Regulus 0.1°N of Moon
28	01:15	Moon at Perigee: 365225 km
29	12:49	Mercury 2.1°S of Moon
30	20:01	NEW MOON
Nov 02	01:14	Jupiter 4.4°S of Moon
05	23	S Taurid Meteor Shower
07	08:34	FIRST QUARTER MOON
08	19:52	Moon at Apogee: 404372 km
10	07:15	Saturn 0.2°S of Moon: Occn.
12	22	N Taurid Meteor Shower
15	08:49	FULL MOON
16	14:53	Aldebaran 0.7°S of Moon
18	05	Leonid Meteor Shower
22	09:28	Regulus 0.1°S of Moon
22	10:22	LAST QUARTER MOON
23	15:48	Moon at Perigee: 370009 km
27	00	Mercury at Superior Conjunction
27	23	Neptune at Opposition
29	08:33	NEW MOON
Dec 06	16:51	Moon at Apogee: 404292 km
07	06:07	FIRST QUARTER MOON
07	15:20	Saturn 0.2°N of Moon: Occn.
09	05	Jupiter in Conjunction with Sun
13	23:35	Aldebaran 0.7°S of Moon
14	18	Geminid Meteor Shower
14	22:41	FULL MOON
17	22	Mars at Opposition
18	16:08	Moon at Perigee: 367712 km
19	14:53	Regulus 0.4°S of Moon
21	16:10	Winter Solstice
21	18:21	LAST QUARTER MOON
23	02	Ursid Meteor Shower
28	23:52	NEW MOON
30	14:47	Mercury 3.9°S of Moon
30	17:28	Venus 3.3°S of Moon

Almanac of Astronomical Events for 2055

Date	GMT (h:m)	Event
Jan 02	21	Mercury 1.0°S of Venus
03	13:48	Moon at Apogee: 404985 km
04	01:47	Saturn 0.6°N of Moon: Occn.
04	08	Quadrantid Meteor Shower
05	12	Earth at Perihelion: 0.98330 AU
05	19	Mercury at Greatest Elong: 19.3°E
06	03:39	FIRST QUARTER MOON
10	09:44	Aldebaran 0.6°S of Moon
13	11:21	FULL MOON
15	09:13	Moon at Perigee: 362045 km
15	22:35	Regulus 0.5°S of Moon
20	03:24	LAST QUARTER MOON
21	21	Mercury at Inferior Conjunction
24	11:31	Jupiter 3.8°S of Moon
27	17:39	NEW MOON
27	17:52	Partial Solar Eclipse; mag=0.693
30	03:04	Venus 0.1°N of Moon: Occn.
31	06:49	Moon at Apogee: 405942 km
31	13:38	Saturn 1.0°N of Moon: Occn.
Feb 04	22:59	FIRST QUARTER MOON
06	19:29	Aldebaran 0.5°S of Moon
11	22:45	Total Lunar Eclipse; mag=1.225
11	22:48	FULL MOON
12	09:04	Regulus 0.5°S of Moon
12	17:59	Moon at Perigee: 357885 km
14	09	Venus 0.9°N of Saturn
15	02	Mercury at Greatest Elong: 26.1°W
18	14:14	LAST QUARTER MOON
21	03:03	Jupiter 3.5°S of Moon
24	07:26	Mercury 1.8°S of Moon
26	12:39	NEW MOON
27	13:56	Moon at Apogee: 406514 km
28	02:08	Saturn 1.2°N of Moon
Mar 01	09:58	Venus 4.0°N of Moon
06	03:22	Aldebaran 0.6°S of Moon
06	14:48	FIRST QUARTER MOON
11	20:29	Regulus 0.5°S of Moon
13	06:25	Moon at Perigee: 356697 km
13	08:57	FULL MOON
20	03:18	LAST QUARTER MOON
20	08	Saturn in Conjunction with Sun
20	15:28	Vernal Equinox
20	16:31	Jupiter 3.1°S of Moon
26	15:17	Moon at Apogee: 406485 km
28	07:01	NEW MOON

Date	GMT (h:m)	Event
Apr 01	08	Mercury at Superior Conjunction
02	09:26	Aldebaran 0.7°S of Moon
02	12	Uranus at Opposition
05	02:43	FIRST QUARTER MOON
08	06:28	Regulus 0.6°S of Moon
09	10:55	Venus 2.3°S of Pleiades
10	17:08	Moon at Perigee: 358662 km
11	17:58	FULL MOON
17	03:43	Jupiter 2.8°S of Moon
18	18:35	LAST QUARTER MOON
23	01:56	Moon at Apogee: 405780 km
23	05	Lyrid Meteor Shower
24	03:46	Saturn 1.7°N of Moon
26	23:17	NEW MOON
27	13	Mercury at Greatest Elong: 20.4°E
29	15:03	Aldebaran 0.9°S of Moon
30	05:31	Mercury 1.3°S of Pleiades
May 03	00:49	Mars 3.6°N of Moon
04	11:10	FIRST QUARTER MOON
05	13:48	Regulus 0.9°S of Moon
05	19	Eta-Aquarid Meteor Shower
08	21:10	Moon at Perigee: 363012 km
11	02:31	FULL MOON
14	12:00	Jupiter 2.7°S of Moon
18	11:30	LAST QUARTER MOON
19	17	Mercury at Inferior Conjunction
20	18:43	Moon at Apogee: 404756 km
21	16:23	Saturn 2.0°N of Moon
21	23	Mars at Aphelion: 1.66600 AU
26	01	Venus at Greatest Elong: 45.4°E
26	12:57	NEW MOON
28	08	Neptune in Conjunction with Sun
29	03:10	Venus 3.8°S of Pollux
30	07:09	Venus 4.3°N of Moon
31	09:37	Mars 1.7°N of Moon
Jun 01	19:15	Regulus 1.1°S of Moon
02	17:01	FIRST QUARTER MOON
05	06:32	Moon at Perigee: 368013 km
09	11:36	FULL MOON
10	16:49	Jupiter 2.8°S of Moon
14	18	Mercury at Greatest Elong: 23.4°W
17	05:02	LAST QUARTER MOON
17	13:16	Moon at Apogee: 404162 km
18	04:02	Saturn 2.2°N of Moon
21	08:39	Summer Solstice
22	08:45	Mercury 2.9°N of Aldebaran
23	05:37	Aldebaran 1.0°S of Moon
23	06:25	Mercury 2.2°N of Moon
24	18	Jupiter at Opposition
25	00:15	NEW MOON
28	01:36	Venus 0.1°S of Moon: Occn.
28	19:08	Mars 0.2°S of Moon: Occn.
29	00:47	Regulus 1.3°S of Moon
30	11:09	Moon at Perigee: 369268 km

Date	GMT (h:m)	Event
Jul 01	21:31	FIRST QUARTER MOON
04	21	Earth at Aphelion: 1.01668 AU
05	00:50	Mars 0.6°N of Regulus
07	18:52	Jupiter 3.1°S of Moon
08	22:11	FULL MOON
15	07	Mercury at Superior Conjunction
15	07:49	Moon at Apogee: 404416 km
15	13:41	Saturn 2.3°N of Moon
16	22:15	LAST QUARTER MOON
20	14:37	Aldebaran 0.9°S of Moon
24	09:48	NEW MOON
24	09:56	Total Solar Eclipse; mag=1.036
26	08:10	Regulus 1.4°S of Moon
27	04:56	Moon at Perigee: 364935 km
27	05:55	Mars 2.1°S of Moon
28	20	Delta-Aquarid Meteor Shower
31	02:11	FIRST QUARTER MOON
Aug 03	11:16	Mercury 0.6°N of Regulus
03	20:29	Jupiter 3.4°S of Moon
04	03	Venus at Inferior Conjunction
07	10:52	Partial Lunar Eclipse; mag=0.959
07	10:57	FULL MOON
11	20:30	Saturn 2.1°N of Moon
12	01:01	Moon at Apogee: 405309 km
13	12	Perseid Meteor Shower
15	14:26	LAST QUARTER MOON
16	23:31	Aldebaran 0.9°S of Moon
22	18:14	NEW MOON
24	04:11	Moon at Perigee: 360218 km
24	18:37	Mars 3.7°S of Moon
25	22	Mercury at Greatest Elong: 27.4°E
26	22	Mercury 3.4°S of Mars
29	08:35	FIRST QUARTER MOON
31	00:43	Jupiter 3.4°S of Moon
Sep 06	01:56	FULL MOON
08	00:25	Saturn 1.9°N of Moon
08	13:59	Moon at Apogee: 406177 km
13	07:15	Aldebaran 1.0°S of Moon
14	05:14	LAST QUARTER MOON
17	22:06	Venus 4.3°S of Moon
19	04:29	Regulus 1.4°S of Moon
21	02:19	NEW MOON
21	12:32	Moon at Perigee: 357315 km
21	15	Mercury at Inferior Conjunction
22	09:43	Mars 4.8°S of Moon
23	00:48	Autumnal Equinox
27	09:53	Jupiter 3.2°S of Moon
27	18:11	FIRST QUARTER MOON
28	09:39	Mars 2.2°N of Spica
29	08	Saturn at Opposition

Date	GMT (h:m)	Event
Oct 05	02:33	Saturn 1.7°N of Moon
05	17:59	Moon at Apogee: 406450 km
05	18:38	FULL MOON
07	06	Mercury at Greatest Elong: 17.9°W
08	11	Uranus in Conjunction with Sun
09	13:28	Venus 1.7°S of Regulus
10	13:34	Aldebaran 1.2°S of Moon
13	18:22	LAST QUARTER MOON
14	10	Venus at Greatest Elong: 46.3°W
16	14:26	Regulus 1.5°S of Moon
17	01:58	Venus 3.1°S of Moon
19	23:43	Moon at Perigee: 357259 km
20	10:49	NEW MOON
22	05	Orionid Meteor Shower
25	00:28	Jupiter 2.7°S of Moon
27	07:53	FIRST QUARTER MOON
Nov 01	04:54	Saturn 1.8°N of Moon
01	20:44	Moon at Apogee: 406106 km
04	12:12	FULL MOON
06	05	S Taurid Meteor Shower
06	15	Mercury at Superior Conjunction
06	19:17	Aldebaran 1.4°S of Moon
12	05:38	LAST QUARTER MOON
12	22:04	Regulus 1.8°S of Moon
13	04	N Taurid Meteor Shower
15	15:29	Venus 3.1°S of Moon
17	09:42	Moon at Perigee: 360295 km
18	11	Leonid Meteor Shower
18	20:34	NEW MOON
21	19:20	Jupiter 2.2°S of Moon
26	01:42	FIRST QUARTER MOON
28	09:37	Saturn 2.0°N of Moon
28	14	Mars in Conjunction with Sun
29	09:39	Moon at Apogee: 405307 km
30	11	Neptune at Opposition
Dec 04	01:45	Aldebaran 1.5°S of Moon
04	05:40	FULL MOON
10	03:37	Regulus 2.0°S of Moon
11	15:05	LAST QUARTER MOON
15	00	Geminid Meteor Shower
15	09:43	Venus 3.0°S of Moon
15	10:52	Moon at Perigee: 365611 km
16	13	Mercury 1.9°S of Jupiter
18	08:15	NEW MOON
19	16:21	Jupiter 1.6°S of Moon
19	22:35	Mercury 2.9°S of Moon
19	23	Mercury at Greatest Elong: 20.3°E
21	21:56	Winter Solstice
23	08	Ursid Meteor Shower
25	17:51	Saturn 2.3°N of Moon
25	22:29	FIRST QUARTER MOON
27	05:00	Moon at Apogee: 404566 km
31	09:44	Aldebaran 1.4°S of Moon

Almanac of Astronomical Events for 2056

Date	GMT (h:m)	Event
Jan 02	22:05	FULL MOON
04	04	Earth at Perihelion: 0.98325 AU
04	14	Quadrantid Meteor Shower
06	00	Mercury at Inferior Conjunction
06	09:14	Regulus 2.2°S of Moon
09	23:13	LAST QUARTER MOON
10	08	Jupiter in Conjunction with Sun
10	16:48	Moon at Perigee: 370190 km
14	08:04	Venus 2.1°S of Moon
15	11:03	Mercury 0.9°N of Moon: Occn.
16	22:10	NEW MOON
16	22:15	Annular Solar Eclipse; mag=0.976
22	05:20	Saturn 2.6°N of Moon
24	02:22	Moon at Apogee: 404486 km
24	20:21	FIRST QUARTER MOON
27	18:42	Aldebaran 1.4°S of Moon
28	11	Mercury at Greatest Elong: 24.8°W
Feb 01	12:24	Pen. Lunar Eclipse; mag=0.906
01	12:36	FULL MOON
02	02	Mars 0.7°S of Jupiter
02	17:02	Regulus 2.2°S of Moon
03	16	Mercury 4.1°N of Venus
04	23:51	Moon at Perigee: 367040 km
08	07:01	LAST QUARTER MOON
08	08	Mercury 0.3°S of Jupiter
12	15	Venus 0.1°N of Jupiter
13	08:05	Jupiter 0.5°S of Moon: Occn.
13	09:32	Venus 0.4°S of Moon: Occn.
13	18	Mercury 0.2°S of Mars
13	20:15	Mars 0.7°S of Moon: Occn.
13	20:25	Mercury 0.9°S of Moon: Occn.
15	14:00	NEW MOON
18	18:35	Saturn 2.7°N of Moon
20	22:27	Moon at Apogee: 405159 km
23	17:11	FIRST QUARTER MOON
24	03:27	Aldebaran 1.5°S of Moon
25	00	Venus 0.4°N of Mars
Mar 01	03:07	Regulus 2.2°S of Moon
02	00:40	FULL MOON
03	19:44	Moon at Perigee: 361481 km
08	15:31	LAST QUARTER MOON
12	00:25	Jupiter 0.0°N of Moon: Occn.
13	20:02	Mars 1.2°N of Moon
14	13:13	Venus 1.9°N of Moon
14	18	Mercury at Superior Conjunction
16	06:52	NEW MOON
19	13:34	Moon at Apogee: 406082 km
19	21:11	Vernal Equinox
22	11:01	Aldebaran 1.7°S of Moon
24	11:17	FIRST QUARTER MOON
28	13:46	Regulus 2.3°S of Moon
31	10:25	FULL MOON

Date	GMT (h:m)	Event
Apr 01	03:57	Moon at Perigee: 357689 km
01	12	Saturn in Conjunction with Sun
06	13	Uranus at Opposition
07	01:33	LAST QUARTER MOON
08	14:24	Jupiter 0.6°N of Moon: Occn.
09	05	Mercury at Greatest Elong: 19.4°E
11	20:38	Mars 2.9°N of Moon
14	23:50	NEW MOON
15	19:35	Moon at Apogee: 406631 km
18	17:20	Aldebaran 1.9°S of Moon
22	12	Lyrid Meteor Shower
23	01:33	FIRST QUARTER MOON
24	23:01	Regulus 2.5°S of Moon
28	23	Mercury at Inferior Conjunction
29	07	Mars at Perihelion: 1.38134 AU
29	14:48	Moon at Perigee: 356811 km
29	18:31	FULL MOON
May 05	01	Eta-Aquarid Meteor Shower
06	02:22	Jupiter 1.1°N of Moon: Occn.
06	13:30	LAST QUARTER MOON
10	22:32	Mars 4.1°N of Moon
11	09:33	Saturn 3.0°N of Moon
12	20:21	Mercury 2.5°N of Moon
12	22:15	Moon at Apogee: 406547 km
14	16:06	NEW MOON
18	11	Mars 1.1°N of Saturn
22	06:02	Regulus 2.7°S of Moon
22	11:50	FIRST QUARTER MOON
26	08	Mercury at Greatest Elong: 25.1°W
26	13	Venus at Superior Conjunction
28	00:00	Moon at Perigee: 358868 km
29	01:58	FULL MOON
29	22	Neptune in Conjunction with Sun
Jun 02	12:13	Jupiter 1.2°N of Moon
05	03:21	LAST QUARTER MOON
07	21:23	Saturn 3.1°N of Moon
09	01:37	Mars 4.4°N of Moon
09	08:50	Moon at Apogee: 405783 km
11	14:32	Mercury 2.3°N of Moon
13	07:04	NEW MOON
18	11:32	Regulus 2.9°S of Moon
20	14:29	Summer Solstice
20	18:48	FIRST QUARTER MOON
25	03:02	Moon at Perigee: 363147 km
27	09:47	FULL MOON
27	10:01	Pen. Lunar Eclipse; mag=0.314
28	17	Mercury at Superior Conjunction
29	19:21	Jupiter 1.1°N of Moon: Occn.

Date	GMT (h:m)	Event
Jul 04	18:55	LAST QUARTER MOON
05	08:26	Saturn 3.1°N of Moon
06	08	Earth at Aphelion: 1.01671 AU
07	00:55	Moon at Apogee: 404772 km
08	04:45	Mars 3.9°N of Moon
09	12:39	Aldebaran 2.0°S of Moon
12	20:20	NEW MOON
12	20:20	Annular Solar Eclipse; mag=0.988
14	05:42	Mercury 0.1°N of Moon: Occn.
15	17:11	Regulus 2.9°S of Moon
19	23:45	FIRST QUARTER MOON
22	11:47	Moon at Perigee: 368056 km
26	18:42	Pen. Lunar Eclipse; mag=0.644
26	18:54	FULL MOON
26	23:31	Jupiter 0.7°N of Moon: Occn.
27	05:09	Mercury 0.2°S of Regulus
28	02	Delta-Aquarid Meteor Shower
29	07	Jupiter at Opposition
Aug 01	18:00	Saturn 2.9°N of Moon
02	18:59	Venus 1.0°N of Regulus
03	11:52	LAST QUARTER MOON
03	19:11	Moon at Apogee: 404218 km
05	20:41	Aldebaran 2.0°S of Moon
06	06:07	Mars 2.8°N of Moon
07	07	Mercury at Greatest Elong: 27.4°E
11	07:48	NEW MOON
12	18	Perseid Meteor Shower
12	21:51	Venus 2.8°S of Moon
16	15:34	Moon at Perigee: 369236 km
18	04:13	FIRST QUARTER MOON
23	01:49	Jupiter 0.4°N of Moon: Occn.
25	06:00	FULL MOON
29	01:12	Saturn 2.6°N of Moon
31	14:03	Moon at Apogee: 404515 km
Sep 02	04:54	Aldebaran 2.2°S of Moon
02	05:43	LAST QUARTER MOON
03	18	Mercury at Inferior Conjunction
04	04:00	Mars 1.6°N of Moon
08	09:30	Regulus 2.9°S of Moon
09	17:47	NEW MOON
11	17:37	Venus 4.9°S of Moon
12	10:34	Moon at Perigee: 364705 km
16	02:04	Venus 2.1°N of Spica
16	09:50	FIRST QUARTER MOON
19	04:38	Jupiter 0.4°N of Moon: Occn.
19	22	Mercury at Greatest Elong: 17.9°W
22	06:40	Autumnal Equinox
23	19:34	FULL MOON
25	05:40	Saturn 2.4°N of Moon
28	07:57	Moon at Apogee: 405424 km
29	12:34	Aldebaran 2.4°S of Moon

Date	GMT (h:m)	Event
Oct 01	23:33	LAST QUARTER MOON
02	20:47	Mars 0.5°N of Moon: Occn.
05	19:29	Regulus 3.0°S of Moon
09	03:00	NEW MOON
10	11:04	Moon at Perigee: 359834 km
11	18	Saturn at Opposition
12	12	Uranus in Conjunction with Sun
15	17:58	FIRST QUARTER MOON
16	10:49	Jupiter 0.7°N of Moon: Occn.
17	07	Mercury at Superior Conjunction
21	11	Orionid Meteor Shower
22	08:04	Saturn 2.4°N of Moon
23	11:46	FULL MOON
24	03:39	Venus 2.9°N of Antares
25	20:58	Moon at Apogee: 406235 km
26	19:19	Aldebaran 2.6°S of Moon
31	06:33	Mars 0.2°S of Moon: Occn.
31	16:12	LAST QUARTER MOON
Nov 02	04:42	Regulus 3.3°S of Moon
07	12:21	NEW MOON
07	20:56	Moon at Perigee: 357013 km
10	11:20	Venus 3.5°S of Moon
12	22:12	Jupiter 1.2°N of Moon: Occn.
14	05:33	FIRST QUARTER MOON
15	04:42	Mercury 2.4°N of Antares
17	17	Leonid Meteor Shower
18	10:18	Saturn 2.6°N of Moon
21	23:18	Moon at Apogee: 406436 km
22	06:12	FULL MOON
23	01:28	Aldebaran 2.7°S of Moon
28	05:55	Mars 0.2°S of Moon: Occn.
29	11:56	Regulus 3.5°S of Moon
30	06:37	LAST QUARTER MOON
Dec 01	21	Mercury at Greatest Elong: 21.4°E
01	22	Neptune at Opposition
06	09:40	Moon at Perigee: 357324 km
06	22:31	NEW MOON
08	08:04	Mercury 2.0°S of Moon
10	07:04	Venus 0.4°S of Moon: Occn.
10	14:52	Jupiter 1.7°N of Moon
13	20:54	FIRST QUARTER MOON
14	06	Geminid Meteor Shower
15	14:52	Saturn 2.9°N of Moon
15	21	Venus 1.5°S of Jupiter
19	02:50	Moon at Apogee: 406059 km
20	06	Mercury at Inferior Conjunction
20	07:47	Aldebaran 2.7°S of Moon
21	03:52	Winter Solstice
22	01:34	FULL MOON
22	01:47	Pen. Lunar Eclipse; mag=0.786
22	15	Ursid Meteor Shower
25	13:52	Mars 0.9°N of Moon: Occn.
26	17:36	Regulus 3.6°S of Moon
29	18:22	LAST QUARTER MOON
31	14	Venus at Greatest Elong: 47.2°E

Almanac of Astronomical Events for 2057

Date	GMT (h:m)	Event
Jan 03	04	Earth at Perihelion: 0.98333 AU
03	19:47	Mercury 0.7°N of Moon: Occn.
03	19:56	Moon at Perigee: 360814 km
03	20	Quadrantid Meteor Shower
05	09:46	Total Solar Eclipse; mag=1.029
05	09:49	NEW MOON
07	11:06	Jupiter 2.3°N of Moon
08	20:46	Venus 4.0°N of Moon
09	20	Mercury at Greatest Elong: 23.3°W
11	23:31	Saturn 3.0°N of Moon
12	15:34	FIRST QUARTER MOON
15	17:52	Moon at Apogee: 405203 km
16	14:51	Aldebaran 2.7°S of Moon
20	20:01	FULL MOON
21	05:35	Mars 2.4°N of Moon
22	23:28	Regulus 3.5°S of Moon
24	00	Mars at Opposition
28	03:44	LAST QUARTER MOON
31	18:59	Moon at Perigee: 366306 km
Feb 02	17:57	Mercury 0.1°N of Moon: Occn.
03	22:10	NEW MOON
08	11:58	Saturn 3.0°N of Moon
11	12:25	FIRST QUARTER MOON
12	12	Jupiter in Conjunction with Sun
12	14:05	Moon at Apogee: 404414 km
12	22:46	Aldebaran 2.8°S of Moon
16	20:31	Mars 2.8°N of Moon
19	07:00	Regulus 3.5°S of Moon
19	11:56	FULL MOON
25	14	Mercury at Superior Conjunction
26	11:30	LAST QUARTER MOON
26	15:47	Moon at Perigee: 370210 km
28	18:41	Mars 3.2°S of Pollux
Mar 04	04:24	Jupiter 3.2°N of Moon
05	11:25	NEW MOON
08	02:26	Saturn 2.9°N of Moon
12	07:00	Aldebaran 3.0°S of Moon
12	10:58	Moon at Apogee: 404319 km
13	09:35	FIRST QUARTER MOON
13	12	Venus at Inferior Conjunction
16	03:28	Mars 1.8°N of Moon
18	16:07	Regulus 3.6°S of Moon
20	03:08	Vernal Equinox
21	00:45	FULL MOON
23	00	Mercury 4.7°N of Saturn
23	06	Mercury at Greatest Elong: 18.6°E
24	08:23	Moon at Perigee: 366570 km
27	18:39	LAST QUARTER MOON
31	22:04	Jupiter 3.6°N of Moon

Date	GMT (h:m)	Event
Apr 04	01:31	NEW MOON
07	21	Mars at Aphelion: 1.66599 AU
08	14:51	Aldebaran 3.3°S of Moon
09	05:48	Moon at Apogee: 404996 km
10	02	Mercury at Inferior Conjunction
11	15	Uranus at Opposition
12	04:59	FIRST QUARTER MOON
13	01:03	Mars 0.2°N of Moon: Occn.
15	01:33	Regulus 3.8°S of Moon
15	03	Saturn in Conjunction with Sun
19	10:49	FULL MOON
21	04:54	Moon at Perigee: 361376 km
22	18	Lyrid Meteor Shower
26	02:06	LAST QUARTER MOON
28	13:15	Jupiter 4.0°N of Moon
May 01	08:57	Mercury 2.7°N of Moon
02	06:25	Saturn 2.7°N of Moon
03	16:32	NEW MOON
05	07	Eta-Aquarid Meteor Shower
05	21:51	Aldebaran 3.4°S of Moon
06	19:52	Moon at Apogee: 405919 km
08	01	Mercury at Greatest Elong: 26.5°W
11	06:26	Mars 1.6°S of Moon
11	21:06	FIRST QUARTER MOON
12	09:57	Regulus 4.0°S of Moon
14	12	Mercury 0.8°S of Saturn
18	19:02	FULL MOON
19	12:10	Moon at Perigee: 357926 km
22	13	Venus at Greatest Elong: 46.0°W
25	10:40	LAST QUARTER MOON
26	02:22	Jupiter 4.2°N of Moon
29	02:09	Venus 3.2°N of Moon
29	18:37	Saturn 2.6°N of Moon
Jun 01	12	Neptune in Conjunction with Sun
02	08:11	NEW MOON
03	02:13	Moon at Apogee: 406459 km
06	20	Venus 0.1°S of Saturn
08	15:24	Mars 3.2°S of Moon
08	16:47	Regulus 4.1°S of Moon
10	09:30	FIRST QUARTER MOON
10	15:36	Mars 0.7°N of Regulus
13	04	Mercury at Superior Conjunction
16	22:08	Moon at Perigee: 357134 km
17	02:18	FULL MOON
17	02:25	Partial Lunar Eclipse; mag=0.755
20	20:19	Summer Solstice
22	13:38	Jupiter 4.2°N of Moon
23	21:08	LAST QUARTER MOON
26	05:49	Saturn 2.4°N of Moon
28	01:17	Venus 0.9°N of Moon: Occn.
29	10:03	Aldebaran 3.4°S of Moon
30	05:32	Moon at Apogee: 406331 km

Date	GMT (h:m)	Event
Jul 01	23:38	Annular Solar Eclipse; mag=0.946
01	23:47	NEW MOON
03	21:04	Mercury 1.1°S of Moon: Occn.
04	17	Earth at Aphelion: 1.01670 AU
05	22:32	Regulus 4.1°S of Moon
07	02:23	Mars 4.4°S of Moon
09	18:37	FIRST QUARTER MOON
11	16:42	Venus 3.2°N of Aldebaran
15	07:01	Moon at Perigee: 359100 km
16	09:28	FULL MOON
19	22:42	Jupiter 4.0°N of Moon
20	11	Mercury at Greatest Elong: 26.8°E
23	10:09	LAST QUARTER MOON
23	16:12	Saturn 2.1°N of Moon
25	09:14	Mercury 1.9°S of Regulus
26	16:29	Aldebaran 3.5°S of Moon
27	15:44	Moon at Apogee: 405562 km
28	06:09	Venus 1.1°S of Moon: Occn.
28	09	Delta-Aquarid Meteor Shower
31	14:32	NEW MOON
Aug 02	04:20	Regulus 4.1°S of Moon
08	01:30	FIRST QUARTER MOON
12	10:32	Moon at Perigee: 363290 km
13	01	Perseid Meteor Shower
14	17:21	FULL MOON
16	05:01	Jupiter 3.8°N of Moon
17	07	Mercury at Inferior Conjunction
20	01:29	Saturn 1.8°N of Moon
22	02:02	LAST QUARTER MOON
22	23:49	Aldebaran 3.6°S of Moon
24	07:38	Moon at Apogee: 404603 km
27	11:59	Venus 2.7°S of Moon
30	03:54	NEW MOON
Sep 03	11	Mercury at Greatest Elong: 18.1°W
04	14	Jupiter at Opposition
06	07:18	FIRST QUARTER MOON
08	05:51	Mars 2.0°N of Spica
08	14:31	Mercury 0.5°N of Regulus
08	19:57	Moon at Perigee: 368286 km
12	08:41	Jupiter 3.6°N of Moon
13	02:53	FULL MOON
16	09:04	Saturn 1.5°N of Moon
17	06:39	Venus 0.5°N of Regulus
19	07:55	Aldebaran 3.9°S of Moon
20	20:25	LAST QUARTER MOON
21	02:28	Moon at Apogee: 404133 km
22	12:23	Autumnal Equinox
25	19:24	Regulus 4.1°S of Moon
26	17:00	Venus 3.8°S of Moon
28	16:00	NEW MOON
29	02	Mercury at Superior Conjunction
30	22:52	Mars 4.3°S of Moon

Date	GMT (h:m)	Event
Oct 03	19:39	Moon at Perigee: 369392 km
05	13:13	FIRST QUARTER MOON
09	11:08	Jupiter 3.6°N of Moon
12	15:01	FULL MOON
13	14:21	Saturn 1.4°N of Moon
17	12	Uranus in Conjunction with Sun
18	22:29	Moon at Apogee: 404515 km
20	16:09	LAST QUARTER MOON
21	17	Orionid Meteor Shower
25	10	Saturn at Opposition
26	21:24	Venus 3.5°S of Moon
28	03:19	NEW MOON
29	09:48	Mercury 4.7°S of Moon
29	18:05	Mars 2.8°S of Moon
30	17:10	Moon at Perigee: 364513 km
Nov 03	20:24	FIRST QUARTER MOON
04	13	Mercury 1.9°S of Mars
05	15:03	Jupiter 3.8°N of Moon
05	17	S Taurid Meteor Shower
09	17:00	Mercury 1.9°N of Antares
09	17:32	Saturn 1.6°N of Moon
11	06:24	FULL MOON
12	17	N Taurid Meteor Shower
13	09:30	Mars 3.9°N of Antares
14	14	Mercury at Greatest Elong: 22.7°E
15	17:25	Moon at Apogee: 405449 km
17	23	Leonid Meteor Shower
19	11:31	LAST QUARTER MOON
26	14:22	NEW MOON
27	20:20	Moon at Perigee: 359527 km
Dec 02	23:13	Jupiter 4.1°N of Moon
03	05:54	FIRST QUARTER MOON
04	09	Neptune at Opposition
04	14	Mercury at Inferior Conjunction
06	20:10	Saturn 1.8°N of Moon
11	00:46	FULL MOON
11	00:52	Partial Lunar Eclipse; mag=0.918
13	06:08	Moon at Apogee: 406205 km
14	12	Geminid Meteor Shower
19	05:02	LAST QUARTER MOON
21	09:42	Winter Solstice
22	21	Ursid Meteor Shower
23	09	Mercury at Greatest Elong: 21.9°W
24	12:34	Mercury 0.8°N of Moon: Occn.
26	01:13	Total Solar Eclipse; mag=1.035
26	01:22	NEW MOON
26	07:44	Moon at Perigee: 356883 km
27	07	Venus at Superior Conjunction
30	13:03	Jupiter 4.3°N of Moon

Almanac of Astronomical Events for 2058

Date	GMT (h:m)	Event
Jan 01	18:30	FIRST QUARTER MOON
03	00:54	Saturn 2.0°N of Moon
04	03	Quadrantid Meteor Shower
05	04	Earth at Perihelion: 0.98333 AU
09	06:58	Moon at Apogee: 406377 km
09	20:39	FULL MOON
17	19:43	LAST QUARTER MOON
23	20	Mars in Conjunction with Sun
23	21:00	Moon at Perigee: 357538 km
24	12:14	NEW MOON
27	07:47	Jupiter 4.4°N of Moon
30	09:50	Saturn 1.9°N of Moon
31	10:28	FIRST QUARTER MOON
Feb 05	12:03	Moon at Apogee: 405978 km
07	15	Mercury at Superior Conjunction
08	15:54	FULL MOON
16	07:16	LAST QUARTER MOON
21	06:06	Moon at Perigee: 361268 km
22	22:56	NEW MOON
24	05:06	Jupiter 4.3°N of Moon
26	22:56	Saturn 1.7°N of Moon
27	19	Mercury 1.3°N of Jupiter
28	21	Venus 0.2°S of Jupiter
Mar 02	05:10	FIRST QUARTER MOON
05	03:58	Moon at Apogee: 405080 km
05	18:23	Pollux 4.0°N of Moon
06	15	Mercury at Greatest Elong: 18.2°E
10	08:52	FULL MOON
17	11	Mars at Perihelion: 1.38127 AU
17	15:56	LAST QUARTER MOON
20	09:04	Vernal Equinox
20	21	Jupiter in Conjunction with Sun
21	01:53	Moon at Perigee: 366648 km
23	01	Mercury at Inferior Conjunction
24	09:50	NEW MOON
25	23:09	Venus 3.7°N of Moon
26	14:18	Saturn 1.3°N of Moon
31	15	Mercury 2.4°N of Mars

Date	GMT (h:m)	Event
Apr 01	01:03	FIRST QUARTER MOON
01	16	Venus 1.9°N of Saturn
01	23:36	Moon at Apogee: 404315 km
02	02:05	Pollux 3.7°N of Moon
08	22:55	FULL MOON
12	19	Mars 0.2°N of Jupiter
15	18:36	Moon at Perigee: 369945 km
15	22:27	LAST QUARTER MOON
16	17	Uranus at Opposition
19	23	Mercury at Greatest Elong: 27.5°W
20	15:17	Mercury 2.7°N of Moon
20	22:18	Jupiter 4.1°N of Moon
21	06:24	Mars 4.1°N of Moon
22	02:18	Venus 3.3°S of Pleiades
22	21:29	NEW MOON
23	00	Lyrid Meteor Shower
25	06:10	Venus 1.9°N of Moon
25	14	Mercury 1.7°S of Jupiter
29	03	Saturn in Conjunction with Sun
29	10:05	Pollux 3.5°N of Moon
29	19:18	Moon at Apogee: 404300 km
30	20:18	FIRST QUARTER MOON
May 05	13	Eta-Aquarid Meteor Shower
06	10	Mercury 1.9°S of Mars
08	10:12	FULL MOON
11	17:49	Moon at Perigee: 366328 km
15	03:58	LAST QUARTER MOON
18	15:42	Jupiter 3.8°N of Moon
20	05:53	Mars 3.0°N of Moon
20	19:41	Saturn 0.8°N of Moon: Occn.
22	10:23	NEW MOON
22	10:38	Partial Solar Eclipse; mag=0.414
25	13:59	Venus 0.3°S of Moon: Occn.
26	17:45	Pollux 3.4°N of Moon
27	13:11	Moon at Apogee: 405059 km
28	16	Mercury at Superior Conjunction
30	13:33	FIRST QUARTER MOON
30	20	Mars 1.7°N of Saturn
Jun 04	02	Neptune in Conjunction with Sun
06	19:14	Total Lunar Eclipse; mag=1.661
06	19:15	FULL MOON
08	13:26	Moon at Perigee: 361344 km
13	09:50	LAST QUARTER MOON
15	06:42	Jupiter 3.5°N of Moon
17	07:49	Saturn 0.5°N of Moon: Occn.
18	05:16	Mars 1.5°N of Moon
21	00:18	Partial Solar Eclipse; mag=0.126
21	00:35	NEW MOON
21	02:03	Summer Solstice
23	00:39	Pollux 3.4°N of Moon
23	03:35	Mercury 1.9°S of Moon
24	03:14	Moon at Apogee: 406033 km
24	17:48	Venus 2.7°S of Moon
29	04:13	FIRST QUARTER MOON

Date	GMT (h:m)	Event
Jul 02	08	Mercury at Greatest Elong: 25.8°E
05	16	Earth at Aphelion: 1.01664 AU
06	02:46	FULL MOON
06	19:45	Moon at Perigee: 357903 km
08	04:44	Venus 0.9°N of Regulus
12	17:28	LAST QUARTER MOON
12	19:29	Jupiter 3.0°N of Moon
14	18:22	Saturn 0.1°N of Moon: Occn.
17	03:36	Mars 0.1°S of Moon: Occn.
20	15:40	NEW MOON
21	10:15	Moon at Apogee: 406575 km
28	15	Delta-Aquarid Meteor Shower
28	16:19	FIRST QUARTER MOON
30	02	Mercury at Inferior Conjunction
Aug 04	05:22	Moon at Perigee: 356996 km
04	09:37	FULL MOON
05	13	Venus at Greatest Elong: 45.8°E
09	06:03	Jupiter 2.6°N of Moon
11	03:53	Saturn 0.3°S of Moon: Occn.
11	04:00	LAST QUARTER MOON
13	07	Perseid Meteor Shower
15	00:17	Mars 1.6°S of Moon
16	12:45	Pollux 3.4°N of Moon
17	13:13	Moon at Apogee: 406424 km
17	17	Mercury at Greatest Elong: 18.7°W
19	07:03	NEW MOON
27	02:10	FIRST QUARTER MOON
Sep 01	14:39	Moon at Perigee: 358916 km
02	16:51	FULL MOON
04	13:53	Venus 2.2°S of Spica
05	14:09	Jupiter 2.3°N of Moon
07	12:41	Saturn 0.7°S of Moon: Occn.
09	18:07	LAST QUARTER MOON
11	18	Mercury at Superior Conjunction
12	19:07	Pollux 3.2°N of Moon
12	19:11	Mars 2.7°S of Moon
13	03	Jupiter at Perihelion: 4.95141 AU
13	22:43	Moon at Apogee: 405674 km
17	22:17	NEW MOON
22	18:07	Autumnal Equinox
25	10:14	FIRST QUARTER MOON
29	18:56	Moon at Perigee: 363283 km

Date	GMT (h:m)	Event
Oct 02	01:36	FULL MOON
02	19:31	Jupiter 2.3°N of Moon
04	20:33	Saturn 0.9°S of Moon: Occn.
09	11:41	LAST QUARTER MOON
10	02:25	Pollux 3.0°N of Moon
11	12:14	Mars 3.5°S of Moon
11	14:58	Moon at Apogee: 404766 km
12	10	Jupiter at Opposition
14	10	Venus at Inferior Conjunction
17	13:05	NEW MOON
19	07:33	Mercury 4.1°S of Moon
21	23	Orionid Meteor Shower
22	12	Uranus in Conjunction with Sun
24	17:16	FIRST QUARTER MOON
27	04:18	Moon at Perigee: 368573 km
28	04	Mercury at Greatest Elong: 24.1°E
29	22:43	Jupiter 2.6°N of Moon
31	12:54	FULL MOON
Nov 01	02:52	Saturn 0.9°S of Moon: Occn.
06	10:36	Pollux 2.8°N of Moon
08	05	Saturn at Opposition
08	07:47	LAST QUARTER MOON
08	10:49	Moon at Apogee: 404362 km
09	02:43	Mars 3.6°S of Moon
15	06:54	Mars 1.4°N of Regulus
16	03:09	NEW MOON
16	03:21	Partial Solar Eclipse; mag=0.764
17	05:34	Antares 4.3°S of Moon
18	05	Leonid Meteor Shower
18	21	Mercury at Inferior Conjunction
20	22:40	Moon at Perigee: 369235 km
23	00:16	FIRST QUARTER MOON
26	01:35	Jupiter 2.9°N of Moon
28	07:15	Saturn 0.6°S of Moon: Occn.
28	20:49	Venus 3.5°N of Spica
30	03:14	Total Lunar Eclipse; mag=1.426
30	03:17	FULL MOON
Dec 03	18:59	Pollux 2.8°N of Moon
06	04	Mercury at Greatest Elong: 20.6°W
06	07:43	Moon at Apogee: 404781 km
06	20	Neptune at Opposition
07	12:20	Mars 3.1°S of Moon
08	04:51	LAST QUARTER MOON
12	00:55	Venus 0.2°S of Moon: Occn.
14	06:15	Mercury 1.3°N of Moon
14	18	Geminid Meteor Shower
15	16:12	NEW MOON
18	01:25	Moon at Perigee: 363889 km
21	15:24	Winter Solstice
22	08:27	FIRST QUARTER MOON
23	03	Ursid Meteor Shower
23	07:01	Jupiter 3.0°N of Moon
25	03	Venus at Greatest Elong: 46.9°W
25	10:41	Saturn 0.4°S of Moon: Occn.
29	20:25	FULL MOON
31	02:41	Pollux 2.9°N of Moon

Almanac of Astronomical Events for 2059

Date	GMT (h:m)	Event
Jan 03	02:24	Moon at Apogee: 405679 km
03	11	Earth at Perihelion: 0.98331 AU
04	09	Quadrantid Meteor Shower
04	12:20	Mars 2.0°S of Moon
07	01:06	LAST QUARTER MOON
10	18:47	Venus 3.1°N of Moon
11	01:51	Antares 4.3°S of Moon
14	03:57	NEW MOON
15	06:49	Moon at Perigee: 358945 km
19	15	Mercury at Superior Conjunction
19	17:34	Jupiter 2.8°N of Moon
20	18:51	FIRST QUARTER MOON
21	15:43	Saturn 0.4°S of Moon: Occn.
27	09:15	Pollux 2.9°N of Moon
28	15:11	FULL MOON
30	13:03	Moon at Apogee: 406368 km
31	19:42	Mars 0.9°S of Moon: Occn.
Feb 05	18:49	LAST QUARTER MOON
07	11:33	Antares 4.0°S of Moon
09	13:48	Venus 4.7°N of Moon
12	14:27	NEW MOON
12	18:34	Moon at Perigee: 356693 km
16	09:45	Jupiter 2.3°N of Moon
18	00:42	Saturn 0.7°S of Moon: Occn.
18	02	Mercury at Greatest Elong: 18.1°E
19	07:57	FIRST QUARTER MOON
23	15:06	Pollux 2.8°N of Moon
23	18	Mars at Aphelion: 1.66602 AU
26	13:16	Moon at Apogee: 406509 km
27	04	Mars at Opposition
27	09:02	Mars 0.7°S of Moon: Occn.
27	10:06	FULL MOON
Mar 05	18	Mercury at Inferior Conjunction
06	18:56	Antares 3.8°S of Moon
07	08:47	LAST QUARTER MOON
11	07:49	Venus 5.0°N of Moon
13	06:38	Moon at Perigee: 357773 km
14	00:05	NEW MOON
16	05:58	Jupiter 1.7°N of Moon
17	13:57	Saturn 1.0°S of Moon: Occn.
18	13:05	Pleiades 4.1°N of Moon
20	14:44	Vernal Equinox
20	23:30	FIRST QUARTER MOON
22	21:19	Pollux 2.5°N of Moon
25	20:15	Moon at Apogee: 406038 km
25	21:28	Mars 1.6°S of Moon
29	03:47	FULL MOON

Date	GMT (h:m)	Event
Apr 02	04	Mercury at Greatest Elong: 27.8°W
03	00:39	Antares 3.5°S of Moon
05	18:46	LAST QUARTER MOON
07	05	Mercury 3.6°N of Venus
10	03:01	Venus 3.8°N of Moon
10	10:16	Mercury 2.4°N of Moon
10	13:47	Moon at Perigee: 361666 km
12	09:28	NEW MOON
14	05:44	Saturn 1.4°S of Moon
14	23:02	Pleiades 4.0°N of Moon
19	04:46	Pollux 2.4°N of Moon
19	16:37	FIRST QUARTER MOON
21	18	Uranus at Opposition
22	02:45	Mars 3.0°S of Moon
22	12:03	Moon at Apogee: 405078 km
23	06	Lyrid Meteor Shower
25	17:56	Mars 1.6°N of Regulus
27	15	Jupiter in Conjunction with Sun
27	19:17	FULL MOON
30	06:27	Antares 3.5°S of Moon
May 05	01:29	LAST QUARTER MOON
05	19	Eta-Aquarid Meteor Shower
08	06:41	Moon at Perigee: 366831 km
10	01:07	Venus 1.2°N of Moon: Occn.
11	19:15	NEW MOON
11	19:20	Total Solar Eclipse; mag=1.024
13	02	Mercury at Superior Conjunction
13	13	Saturn in Conjunction with Sun
16	13:16	Pollux 2.3°N of Moon
19	10:22	FIRST QUARTER MOON
20	00:24	Mars 4.0°S of Moon
20	06:41	Moon at Apogee: 404327 km
24	01	Venus 0.4°S of Jupiter
27	07:54	Partial Lunar Eclipse; mag=0.183
27	08:04	FULL MOON
27	13:44	Antares 3.5°S of Moon
Jun 02	02	Venus 0.9°N of Saturn
02	22:39	Moon at Perigee: 369676 km
03	06:20	LAST QUARTER MOON
06	17	Neptune in Conjunction with Sun
07	20:25	Jupiter 0.1°S of Moon: Occn.
08	12:11	Saturn 2.0°S of Moon
08	17:20	Pleiades 4.0°N of Moon
10	05:57	NEW MOON
12	05:51	Mercury 2.6°S of Moon
12	21:53	Pollux 2.3°N of Moon
13	23	Mercury at Greatest Elong: 24.3°E
17	01:27	Moon at Apogee: 404358 km
17	08:23	Mars 4.4°S of Moon
18	03:55	FIRST QUARTER MOON
21	07:47	Summer Solstice
23	22:37	Antares 3.4°S of Moon
25	18:12	FULL MOON
29	00:49	Moon at Perigee: 366101 km

Date	GMT (h:m)	Event
Jul 02	10:53	LAST QUARTER MOON
05	13:04	Jupiter 0.8°S of Moon: Occn.
05	23:51	Pleiades 3.9°N of Moon
06	00:19	Saturn 2.4°S of Moon
06	22	Earth at Aphelion: 1.01672 AU
09	17:58	NEW MOON
11	01	Mercury at Inferior Conjunction
14	18:59	Moon at Apogee: 405149 km
15	22:42	Mars 4.1°S of Moon
17	20:34	FIRST QUARTER MOON
21	08:13	Antares 3.3°S of Moon
25	02:24	FULL MOON
26	20:05	Moon at Perigee: 361174 km
28	21	Delta-Aquarid Meteor Shower
31	14	Mercury at Greatest Elong: 19.6°W
31	16:43	LAST QUARTER MOON
Aug 02	02:59	Jupiter 1.4°S of Moon
02	05:21	Pleiades 3.7°N of Moon
02	10:15	Saturn 2.8°S of Moon
03	01	Venus at Superior Conjunction
06	12:06	Pollux 2.4°N of Moon
08	07:37	NEW MOON
11	09:12	Moon at Apogee: 406129 km
13	13	Perseid Meteor Shower
13	17:08	Mars 3.0°S of Moon
16	11:41	FIRST QUARTER MOON
17	17:13	Antares 3.1°S of Moon
18	01:06	Mars 1.7°N of Spica
23	09:42	FULL MOON
24	02:14	Moon at Perigee: 357713 km
26	02	Mercury at Superior Conjunction
29	11:29	Pleiades 3.4°N of Moon
29	14:24	Jupiter 2.0°S of Moon
29	18:48	Saturn 3.3°S of Moon
30	01:05	LAST QUARTER MOON
Sep 02	17:43	Pollux 2.2°N of Moon
06	23:01	NEW MOON
07	16:10	Moon at Apogee: 406622 km
11	14:02	Mars 1.5°S of Moon
14	00:35	Antares 2.8°S of Moon
15	00:44	FIRST QUARTER MOON
21	12:34	Moon at Perigee: 356861 km
21	17:18	FULL MOON
23	00:03	Autumnal Equinox
25	08	Jupiter 2.0°N of Saturn
25	10:10	Mercury 0.7°N of Spica
25	19:38	Pleiades 3.2°N of Moon
25	23:38	Jupiter 2.4°S of Moon
26	02:55	Saturn 3.6°S of Moon
28	12:54	LAST QUARTER MOON
29	23:38	Pollux 2.0°N of Moon

Date	GMT (h:m)	Event
Oct 01	02:40	Venus 2.6°N of Spica
04	18:25	Moon at Apogee: 406424 km
06	15:50	NEW MOON
08	06:06	Venus 0.9°S of Moon: Occn.
08	20:08	Mercury 3.7°S of Moon
10	11:54	Mars 0.3°N of Moon: Occn.
10	16	Mercury at Greatest Elong: 25.3°E
11	06:28	Antares 2.7°S of Moon
14	11:38	FIRST QUARTER MOON
19	23:09	Moon at Perigee: 359009 km
21	02:15	FULL MOON
21	21	Mercury 3.1°S of Venus
22	05	Orionid Meteor Shower
23	05:47	Pleiades 3.1°N of Moon
23	06:43	Jupiter 2.3°S of Moon
23	10:55	Saturn 3.7°S of Moon
24	14:51	Mars 3.6°N of Antares
27	06:58	Pollux 1.9°N of Moon
27	11	Uranus in Conjunction with Sun
28	04:32	LAST QUARTER MOON
Nov 01	04:40	Moon at Apogee: 405658 km
03	01	Mercury at Inferior Conjunction
05	09:11	NEW MOON
05	09:16	Annular Solar Eclipse; mag=0.942
07	04:48	Venus 3.7°N of Antares
07	12:09	Antares 2.6°S of Moon
07	14:28	Venus 1.3°N of Moon
08	09:41	Mars 1.9°N of Moon
12	20:45	FIRST QUARTER MOON
17	04:17	Moon at Perigee: 363733 km
18	06	Jupiter at Opposition
18	11	Leonid Meteor Shower
19	08	Mercury at Greatest Elong: 19.5°W
19	11:41	Jupiter 2.0°S of Moon
19	13:00	Partial Lunar Eclipse; mag=0.208
19	13:09	FULL MOON
19	16:31	Pleiades 3.1°N of Moon
19	18:13	Saturn 3.5°S of Moon
22	05	Saturn at Opposition
23	15:59	Pollux 1.9°N of Moon
26	23:42	LAST QUARTER MOON
27	11	Venus 0.4°S of Mars
28	22:30	Moon at Apogee: 404737 km
Dec 05	01:49	NEW MOON
07	07:08	Mars 3.2°N of Moon
07	15:57	Venus 2.8°N of Moon
09	07	Neptune at Opposition
12	04:50	FIRST QUARTER MOON
14	10:07	Moon at Perigee: 369175 km
15	01	Geminid Meteor Shower
16	15:10	Jupiter 1.7°S of Moon
16	23:58	Saturn 3.3°S of Moon
17	01:50	Pleiades 3.1°N of Moon
19	02:11	FULL MOON
21	01:38	Pollux 2.0°N of Moon
21	21:18	Winter Solstice
23	09	Ursid Meteor Shower
26	19:36	Moon at Apogee: 404322 km
26	21:17	LAST QUARTER MOON
30	13	Mercury at Superior Conjunction

Almanac of Astronomical Events for 2060

Date	GMT (h:m)	Event
Jan 01	03:58	Antares 2.6°S of Moon
03	16:40	NEW MOON
04	15	Quadrantid Meteor Shower
04	23	Earth at Perihelion: 0.98335 AU
05	04:47	Mars 4.0°N of Moon
06	12:33	Venus 3.5°N of Moon
08	01:42	Moon at Perigee: 368882 km
10	12:52	FIRST QUARTER MOON
12	19:08	Jupiter 1.6°S of Moon
13	04:29	Saturn 3.2°S of Moon
13	08:42	Pleiades 2.9°N of Moon
17	10:16	Pollux 2.0°N of Moon
17	17:14	FULL MOON
23	16:37	Moon at Apogee: 404707 km
24	23	Mercury 0.1°S of Mars
25	19:14	LAST QUARTER MOON
28	13:30	Antares 2.4°S of Moon
Feb 01	15	Mercury at Greatest Elong: 18.4°E
02	05:22	NEW MOON
02	08	Mars at Perihelion: 1.38146 AU
04	10:39	Moon at Perigee: 363348 km
05	06:32	Venus 3.4°N of Moon
08	21:41	FIRST QUARTER MOON
09	02:20	Jupiter 1.8°S of Moon
09	09:52	Saturn 3.4°S of Moon
09	14:10	Pleiades 2.7°N of Moon
13	17:01	Pollux 2.0°N of Moon
16	09:56	FULL MOON
17	01	Mercury at Inferior Conjunction
20	10:06	Moon at Apogee: 405550 km
21	00:00	Spica 3.9°S of Moon
24	15:06	LAST QUARTER MOON
24	22:14	Antares 2.1°S of Moon
Mar 02	16:11	NEW MOON
03	16:59	Moon at Perigee: 358816 km
05	21:23	Venus 3.0°N of Moon
07	14:38	Jupiter 2.2°S of Moon
07	18:39	Saturn 3.6°S of Moon
07	20:36	Pleiades 2.4°N of Moon
09	07:52	FIRST QUARTER MOON
11	22:31	Pollux 1.8°N of Moon
13	02	Venus at Greatest Elong: 46.3°E
14	12	Mercury at Greatest Elong: 27.6°W
17	03:41	FULL MOON
18	18:59	Moon at Apogee: 406197 km
19	06:40	Spica 3.8°S of Moon
19	20:37	Vernal Equinox
23	05:22	Antares 1.9°S of Moon
25	07:08	LAST QUARTER MOON
30	09:26	Mercury 1.9°N of Moon
31	09	Jupiter 1.1°N of Saturn
Apr 01	01:37	NEW MOON
01	04:11	Moon at Perigee: 357029 km
04	03:52	Venus 2.8°N of Moon
04	05:26	Pleiades 2.2°N of Moon
04	07:33	Saturn 3.9°S of Moon
04	07:46	Jupiter 2.7°S of Moon
05	04:17	Venus 0.5°N of Pleiades
06	03	Mars in Conjunction with Sun
07	19:42	FIRST QUARTER MOON
08	04:33	Pollux 1.6°N of Moon
14	20:02	Moon at Apogee: 406305 km
15	12:48	Spica 3.8°S of Moon
15	21:21	FULL MOON
15	21:35	Pen. Lunar Eclipse; mag=0.767
19	11:19	Antares 1.8°S of Moon
22	12	Lyrid Meteor Shower
23	18:53	LAST QUARTER MOON
25	19	Uranus at Opposition
26	08	Mercury at Superior Conjunction
29	14:53	Moon at Perigee: 358297 km
30	10:08	Total Solar Eclipse; mag=1.066
30	10:11	NEW MOON
May 01	15:59	Pleiades 2.2°N of Moon
02	03:52	Jupiter 3.1°S of Moon
02	10:24	Venus 2.4°N of Moon
05	01	Eta-Aquarid Meteor Shower
05	12:21	Pollux 1.6°N of Moon
07	09:19	FIRST QUARTER MOON
11	14	Mercury 2.6°N of Jupiter
12	03	Mercury 2.2°S of Venus
12	04:06	Moon at Apogee: 405764 km
12	19:08	Spica 3.8°S of Moon
15	13:39	FULL MOON
16	17:13	Antares 1.9°S of Moon
23	03:01	LAST QUARTER MOON
23	03	Venus at Inferior Conjunction
25	13	Mercury at Greatest Elong: 22.7°E
27	07	Saturn in Conjunction with Sun
27	20:38	Moon at Perigee: 362061 km
29	18:23	NEW MOON
31	08:42	Mercury 3.1°S of Moon
Jun 01	21:43	Pollux 1.7°N of Moon
03	07	Jupiter in Conjunction with Sun
06	00:44	FIRST QUARTER MOON
08	19:26	Moon at Apogee: 404799 km
09	02:13	Spica 3.7°S of Moon
13	00:00	Antares 1.9°S of Moon
14	03:37	FULL MOON
15	05	Mars 1.8°N of Saturn
20	06	Mercury at Inferior Conjunction
20	13:44	Summer Solstice
21	08:44	LAST QUARTER MOON
23	15:32	Saturn 3.8°N of Aldebaran
24	12:41	Moon at Perigee: 366989 km
25	11:16	Pleiades 2.2°N of Moon
26	16:47	Mars 3.0°S of Moon
26	20:36	Jupiter 3.9°S of Moon
28	02:58	NEW MOON
29	07:19	Pollux 1.8°N of Moon

Date	GMT (h:m)	Event
Jul 01	10	Mars 0.8°N of Jupiter
04	00	Earth at Aphelion: 1.01668 AU
05	17:38	FIRST QUARTER MOON
06	10:04	Spica 3.5°S of Moon
06	13:25	Moon at Apogee: 404112 km
10	07:57	Antares 1.8°S of Moon
12	23	Mercury 4.2°S of Mars
13	00	Mercury at Greatest Elong: 20.9°W
13	15:08	FULL MOON
15	22:30	Venus 1.4°N of Aldebaran
19	18	Venus 2.3°S of Saturn
20	04:58	Moon at Perigee: 369731 km
20	13:24	LAST QUARTER MOON
22	17:56	Pleiades 2.0°N of Moon
25	09:10	Mars 3.9°S of Moon
27	12:49	NEW MOON
28	03	Delta-Aquarid Meteor Shower
Aug 01	08	Venus at Greatest Elong: 45.7°W
02	18:14	Spica 3.2°S of Moon
03	08:10	Moon at Apogee: 404228 km
04	11:16	FIRST QUARTER MOON
04	16	Venus 2.9°S of Jupiter
06	16:31	Antares 1.6°S of Moon
08	21	Mercury at Superior Conjunction
12	00:51	FULL MOON
12	19	Perseid Meteor Shower
15	06:57	Moon at Perigee: 366169 km
18	18:23	LAST QUARTER MOON
18	23:24	Pleiades 1.7°N of Moon
22	22:32	Pollux 1.7°N of Moon
23	00:04	Mars 4.2°S of Moon
26	00:56	NEW MOON
27	09:16	Mercury 2.9°S of Moon
30	02:01	Spica 2.9°S of Moon
31	02:25	Moon at Apogee: 405090 km
Sep 03	00:49	Antares 1.3°S of Moon
03	04:36	FIRST QUARTER MOON
10	09:44	FULL MOON
11	09	Venus 1.8°S of Mars
12	02:40	Moon at Perigee: 361154 km
15	05:36	Pleiades 1.5°N of Moon
17	01:00	LAST QUARTER MOON
19	04:01	Pollux 1.5°N of Moon
19	19:37	Mercury 0.4°S of Spica
20	14:24	Mars 4.1°S of Moon
22	04	Mercury at Greatest Elong: 26.4°E
22	05:47	Autumnal Equinox
24	15:53	NEW MOON
26	08:58	Spica 2.8°S of Moon
26	21:29	Mercury 3.5°S of Moon
27	17:42	Moon at Apogee: 406086 km
30	08:05	Antares 1.2°S of Moon
30	11:31	Venus 0.0°N of Regulus

Date	GMT (h:m)	Event
Oct 02	20:41	FIRST QUARTER MOON
09	18:41	FULL MOON
09	18:52	Pen. Lunar Eclipse; mag=0.880
10	10:18	Moon at Perigee: 357605 km
12	14:05	Pleiades 1.4°N of Moon
16	09:52	Pollux 1.4°N of Moon
16	10:30	LAST QUARTER MOON
17	02	Mercury at Inferior Conjunction
17	22:37	Mars 0.9°N of Regulus
19	04:51	Mars 3.4°S of Moon
21	05:02	Venus 1.8°S of Moon
21	11	Orionid Meteor Shower
24	09:22	Annular Solar Eclipse; mag=0.928
24	09:25	NEW MOON
25	00:25	Moon at Apogee: 406530 km
27	14:17	Antares 1.2°S of Moon
31	09	Uranus in Conjunction with Sun
Nov 01	10:56	FIRST QUARTER MOON
01	18	Mercury at Greatest Elong: 18.7°W
07	22:11	Moon at Perigee: 356812 km
08	04:02	Pen. Lunar Eclipse; mag=0.027
08	04:17	FULL MOON
09	00:52	Pleiades 1.5°N of Moon
12	17:42	Pollux 1.5°N of Moon
14	00:41	Venus 3.5°N of Spica
14	23:48	LAST QUARTER MOON
16	19:29	Mars 2.0°S of Moon
17	18	Leonid Meteor Shower
19	21:21	Spica 2.8°S of Moon
20	17:57	Venus 2.2°N of Moon
21	01:49	Moon at Apogee: 406318 km
23	04:16	NEW MOON
30	23:10	FIRST QUARTER MOON
Dec 05	08	Saturn at Opposition
06	09:57	Moon at Perigee: 359223 km
06	12:12	Pleiades 1.5°N of Moon
07	14:48	FULL MOON
08	17	Mercury at Superior Conjunction
10	03:46	Pollux 1.6°N of Moon
10	18	Neptune at Opposition
12	17:20	Regulus 4.2°S of Moon
14	07	Geminid Meteor Shower
14	17:15	LAST QUARTER MOON
15	09:04	Mars 0.2°S of Moon: Occn.
17	04:09	Spica 2.7°S of Moon
18	13:24	Moon at Apogee: 405552 km
21	02:51	Antares 1.2°S of Moon
21	03:00	Winter Solstice
21	05:24	Venus 4.4°N of Moon
22	01	Jupiter at Opposition
22	15	Ursid Meteor Shower
22	22:39	NEW MOON
30	09:29	FIRST QUARTER MOON

Almanac of Astronomical Events for 2061

Date	GMT (h:m)	Event
Jan 02	21:48	Pleiades 1.3°N of Moon
03	14:50	Moon at Perigee: 364234 km
03	21	Quadrantid Meteor Shower
04	07	Earth at Perihelion: 0.98330 AU
04	22:20	Jupiter 4.7°S of Moon
06	02:24	FULL MOON
06	14:32	Pollux 1.7°N of Moon
09	03:11	Regulus 4.0°S of Moon
10	21	Mars at Aphelion: 1.66612 AU
12	18:41	Mars 1.7°N of Moon
13	12:05	Spica 2.4°S of Moon
13	13:57	LAST QUARTER MOON
15	02	Mercury at Greatest Elong: 18.9°E
15	08:29	Moon at Apogee: 404636 km
17	10:34	Antares 1.0°S of Moon
21	15:16	NEW MOON
21	16:24	Saturn 3.8°N of Aldebaran
28	18:10	FIRST QUARTER MOON
30	04:39	Pleiades 1.1°N of Moon
30	14:53	Moon at Perigee: 369564 km
30	19	Mercury at Inferior Conjunction
Feb 01	02:21	Jupiter 4.7°S of Moon
02	23:49	Pollux 1.7°N of Moon
04	15:22	FULL MOON
05	13:00	Regulus 3.8°S of Moon
09	19:04	Mars 3.1°N of Moon
09	20:45	Spica 2.1°S of Moon
12	05:47	Moon at Apogee: 404249 km
12	11:52	LAST QUARTER MOON
13	18:56	Antares 0.8°S of Moon
20	05:31	NEW MOON
24	09:19	Moon at Perigee: 368440 km
24	22	Mercury at Greatest Elong: 26.8°W
26	10:04	Pleiades 0.9°N of Moon
27	01:51	FIRST QUARTER MOON
28	07:31	Jupiter 4.8°S of Moon
Mar 02	06:39	Pollux 1.6°N of Moon
04	21:09	Regulus 3.9°S of Moon
06	05:54	FULL MOON
09	03:14	Mars 3.6°N of Moon
09	05:13	Spica 1.9°S of Moon
12	02:00	Moon at Apogee: 404673 km
13	03:08	Antares 0.6°S of Moon
13	23	Venus at Superior Conjunction
14	08:31	LAST QUARTER MOON
20	02:26	Vernal Equinox
20	07:29	Mercury 1.1°N of Moon: Occn.
21	17:23	NEW MOON
23	21:11	Moon at Perigee: 363081 km
25	16:36	Pleiades 0.7°N of Moon
28	09:26	FIRST QUARTER MOON
29	12:03	Pollux 1.4°N of Moon
Apr 01	03:22	Regulus 3.9°S of Moon
02	12	Mars at Opposition
04	17:19	Mars 2.4°N of Moon

Date	GMT (h:m)	Event
Apr 04	21:47	FULL MOON
04	21:52	Total Lunar Eclipse; mag=1.034
05	12:38	Spica 1.9°S of Moon
08	18:12	Moon at Apogee: 405543 km
09	10:30	Antares 0.5°S of Moon
10	09	Mercury at Superior Conjunction
13	02:10	LAST QUARTER MOON
20	02:55	Total Solar Eclipse; mag=1.048
20	03:04	NEW MOON
21	03:02	Moon at Perigee: 358852 km
22	01:32	Pleiades 0.7°N of Moon
22	18	Lyrid Meteor Shower
25	18:11	Pollux 1.5°N of Moon
26	17:55	FIRST QUARTER MOON
28	08:53	Regulus 3.9°S of Moon
30	13:06	Mercury 1.7°S of Pleiades
30	20	Uranus at Opposition
May 01	05:55	Mars 0.4°N of Moon: Occn.
02	18:55	Spica 1.9°S of Moon
04	14:13	FULL MOON
05	08	Eta-Aquarid Meteor Shower
06	02:42	Moon at Apogee: 406209 km
06	16:57	Antares 0.6°S of Moon
07	11	Mercury at Greatest Elong: 21.2°E
12	16:10	LAST QUARTER MOON
18	02	Mercury 0.8°N of Venus
19	11:03	NEW MOON
19	12:47	Moon at Perigee: 357187 km
20	17	Venus 1.9°N of Saturn
20	16:54	Venus 3.8°S of Moon
21	23:15	Jupiter 5.0°S of Moon
23	02:25	Pollux 1.6°N of Moon
25	15:26	Regulus 3.8°S of Moon
26	04:12	FIRST QUARTER MOON
28	11:19	Mars 0.7°S of Moon: Occn.
30	00:49	Spica 1.8°S of Moon
31	00	Mercury at Inferior Conjunction
Jun 02	04:48	Moon at Apogee: 406301 km
02	22:59	Antares 0.6°S of Moon
03	06:09	FULL MOON
08	21	Venus 1.2°N of Jupiter
11	02:42	LAST QUARTER MOON
11	07	Saturn in Conjunction with Sun
15	22:47	Pleiades 0.7°N of Moon
16	22:16	Moon at Perigee: 358371 km
17	18:03	NEW MOON
19	12:02	Venus 3.5°S of Moon
19	12:26	Pollux 1.7°N of Moon
20	19:33	Summer Solstice
21	23:51	Regulus 3.5°S of Moon
22	06:48	Mercury 1.7°N of Aldebaran
24	16:54	FIRST QUARTER MOON
24	23	Mercury at Greatest Elong: 22.4°W
25	10:04	Mars 0.6°S of Moon: Occn.
26	07:19	Spica 1.6°S of Moon
29	12:19	Moon at Apogee: 405739 km
30	05:15	Antares 0.6°S of Moon

Date	GMT (h:m)	Event
Jul 02	20:52	FULL MOON
04	10	Mercury 0.4°S of Saturn
06	08	Earth at Aphelion: 1.01663 AU
09	10	Jupiter in Conjunction with Sun
10	10:23	LAST QUARTER MOON
13	07:44	Pleiades 0.6°N of Moon
15	03:23	Moon at Perigee: 362013 km
17	01:10	NEW MOON
19	09:14	Venus 1.0°N of Regulus
19	09:35	Regulus 3.3°S of Moon
19	10:27	Venus 2.1°S of Moon
19	17:45	Mars 1.3°N of Spica
23	15:00	Spica 1.3°S of Moon
23	20:33	Mars 0.4°N of Moon: Occn.
24	00	Mercury at Superior Conjunction
24	08:05	FIRST QUARTER MOON
27	02:41	Moon at Apogee: 404826 km
27	12:16	Antares 0.4°S of Moon
28	09	Delta-Aquarid Meteor Shower
Aug 01	10:11	FULL MOON
08	16:09	LAST QUARTER MOON
09	14:28	Pleiades 0.3°N of Moon
11	19:47	Moon at Perigee: 366956 km
13	01	Perseid Meteor Shower
13	07:49	Pollux 1.7°N of Moon
15	09:39	NEW MOON
17	00:05	Mercury 2.1°S of Moon
18	13:12	Venus 0.2°S of Moon: Occn.
19	23:37	Spica 1.1°S of Moon
21	14:20	Mars 1.7°N of Moon
23	01:18	FIRST QUARTER MOON
23	20:03	Antares 0.2°S of Moon
23	20:28	Moon at Apogee: 404229 km
30	22:18	FULL MOON
Sep 03	13:51	Venus 1.3°N of Spica
04	16	Mercury at Greatest Elong: 27.1°E
05	19:56	Pleiades 0.2°N of Moon
06	10:37	Moon at Perigee: 369805 km
06	21:12	LAST QUARTER MOON
09	14:47	Pollux 1.6°N of Moon
12	03:37	Regulus 3.3°S of Moon
13	20:37	NEW MOON
15	16:42	Mercury 3.7°S of Moon
16	08:20	Spica 1.0°S of Moon
17	16:47	Venus 0.9°N of Moon: Occn.
19	12:22	Mars 2.8°N of Moon
20	04:07	Antares 0.1°S of Moon
20	15:48	Moon at Apogee: 404443 km
21	19:44	FIRST QUARTER MOON
22	11:31	Autumnal Equinox
29	09:32	FULL MOON
29	09:36	Total Lunar Eclipse; mag=1.162
30	20	Mercury at Inferior Conjunction
Oct 01	16:30	Mars 3.2°N of Antares
02	13:00	Moon at Perigee: 365993 km
03	02:07	Pleiades 0.1°N of Moon

Date	GMT (h:m)	Event
Oct 06	02:57	LAST QUARTER MOON
06	20:15	Pollux 1.6°N of Moon
09	10:05	Regulus 3.3°S of Moon
12	00:10	Mercury 0.3°N of Moon: Occn.
13	10:30	Annular Solar Eclipse; mag=0.947
13	10:41	NEW MOON
15	21:21	Venus 1.0°N of Antares
16	08	Mercury at Greatest Elong: 18.1°W
17	11:50	Antares 0.1°S of Moon
17	15:46	Venus 0.9°N of Moon: Occn.
17	18	Venus at Greatest Elong: 46.9°E
18	11:00	Moon at Apogee: 405366 km
18	12:41	Mars 3.6°N of Moon
21	14:24	FIRST QUARTER MOON
21	18	Orionid Meteor Shower
28	20:12	FULL MOON
30	10:28	Moon at Perigee: 360686 km
30	10:42	Pleiades 0.1°N of Moon
Nov 03	02:12	Pollux 1.6°N of Moon
04	10:53	LAST QUARTER MOON
05	06	Uranus in Conjunction with Sun
05	15:34	Regulus 3.2°S of Moon
09	22:45	Spica 1.0°S of Moon
12	03:40	NEW MOON
13	18:45	Antares 0.1°S of Moon
15	02:30	Moon at Apogee: 406329 km
16	00:49	Venus 1.1°N of Moon: Occn.
16	14:14	Mars 3.7°N of Moon
17	22	Mercury at Superior Conjunction
18	00	Leonid Meteor Shower
20	08:11	FIRST QUARTER MOON
26	21:36	Pleiades 0.2°N of Moon
27	06:32	FULL MOON
27	19:57	Moon at Perigee: 357101 km
30	10:32	Pollux 1.8°N of Moon
Dec 02	22:03	Regulus 3.0°S of Moon
03	22:12	LAST QUARTER MOON
07	04:33	Spica 0.8°S of Moon
11	22:32	NEW MOON
12	07:27	Moon at Apogee: 406709 km
13	04	Neptune at Opposition
14	13	Geminid Meteor Shower
15	16:36	Mars 3.3°N of Moon
17	06	Mercury 2.7°S of Venus
19	11	Saturn at Opposition
19	23:58	FIRST QUARTER MOON
20	05	Mars at Perihelion: 1.38120 AU
21	08:49	Winter Solstice
22	21	Ursid Meteor Shower
24	08:56	Pleiades 0.1°N of Moon
26	08:56	Moon at Perigee: 356616 km
26	16:53	FULL MOON
27	14	Venus at Inferior Conjunction
27	21:21	Pollux 2.0°N of Moon
28	17:41	Jupiter 3.9°S of Moon
29	08	Mercury at Greatest Elong: 19.7°E
30	07:00	Regulus 2.7°S of Moon

Almanac of Astronomical Events for 2062

Date	GMT (h:m)	Event
Jan 02	13:21	LAST QUARTER MOON
02	21	Earth at Perihelion: 0.98334 AU
03	10:56	Spica 0.5°S of Moon
04	03	Quadrantid Meteor Shower
07	07:05	Antares 0.1°S of Moon
08	09:05	Moon at Apogee: 406473 km
10	17:52	NEW MOON
13	19:32	Mars 2.0°N of Moon
14	19	Mercury at Inferior Conjunction
18	12:51	FIRST QUARTER MOON
20	18:23	Pleiades 0.1°S of Moon
23	17	Jupiter at Opposition
23	20:23	Moon at Perigee: 359491 km
24	08:43	Pollux 2.0°N of Moon
24	23:45	Jupiter 3.9°S of Moon
25	03:37	FULL MOON
26	17:52	Regulus 2.5°S of Moon
30	19:03	Spica 0.2°S of Moon
Feb 01	07:43	LAST QUARTER MOON
03	13:53	Antares 0.1°N of Moon
04	22:20	Moon at Apogee: 405649 km
07	07	Mercury at Greatest Elong: 25.6°W
09	12:11	NEW MOON
11	22:10	Mars 0.2°N of Moon: Occn.
16	22:38	FIRST QUARTER MOON
17	01:04	Pleiades 0.3°S of Moon
20	18:18	Pollux 1.9°N of Moon
20	23:11	Moon at Perigee: 364741 km
21	04:29	Jupiter 4.0°S of Moon
23	04:32	Regulus 2.5°S of Moon
23	15:08	FULL MOON
27	04:35	Spica 0.1°S of Moon
Mar 02	21:44	Antares 0.3°N of Moon
03	03:49	LAST QUARTER MOON
04	17:42	Moon at Apogee: 404686 km
08	04	Venus at Greatest Elong: 46.7°W
11	04:13	NEW MOON
11	04:24	Partial Solar Eclipse; mag=0.933
12	23:13	Mars 1.8°S of Moon
16	06:28	Pleiades 0.4°S of Moon
18	05:58	FIRST QUARTER MOON
19	16:28	Moon at Perigee: 369624 km
20	01:10	Pollux 1.8°N of Moon
20	08:07	Vernal Equinox
20	09:04	Jupiter 4.1°S of Moon
22	13:07	Regulus 2.5°S of Moon
25	01	Mercury at Superior Conjunction
25	03:32	Total Lunar Eclipse; mag=1.270
25	03:35	FULL MOON
26	14:06	Spica 0.0°S of Moon
30	06:10	Antares 0.3°N of Moon

Date	GMT (h:m)	Event
Apr 01	14:06	Moon at Apogee: 404300 km
01	23:55	LAST QUARTER MOON
05	23:32	Venus 2.9°N of Moon
09	17:17	NEW MOON
11	00:50	Mercury 1.6°S of Moon
12	12:56	Pleiades 0.4°S of Moon
13	17:32	Moon at Perigee: 367906 km
16	06:34	Pollux 1.9°N of Moon
16	12:03	FIRST QUARTER MOON
16	15:46	Jupiter 4.1°S of Moon
18	19:24	Regulus 2.5°S of Moon
19	19	Mercury at Greatest Elong: 19.9°E
22	22:13	Spica 0.1°S of Moon
23	01	Lyrid Meteor Shower
23	16:57	FULL MOON
26	14:19	Antares 0.3°N of Moon
29	08:55	Moon at Apogee: 404747 km
May 01	18:33	LAST QUARTER MOON
05	14	Eta-Aquarid Meteor Shower
05	20	Uranus at Opposition
06	04:32	Venus 2.4°S of Moon
09	03:22	NEW MOON
10	22	Mercury at Inferior Conjunction
11	06:10	Moon at Perigee: 362785 km
13	12:48	Pollux 2.0°N of Moon
14	02:30	Jupiter 3.8°S of Moon
15	18:17	FIRST QUARTER MOON
16	00:54	Regulus 2.3°S of Moon
20	04:38	Spica 0.0°N of Moon
23	07:03	FULL MOON
23	21:28	Antares 0.2°N of Moon
27	00:01	Moon at Apogee: 405628 km
31	10:44	LAST QUARTER MOON
Jun 05	15	Saturn at Perihelion: 9.03096 AU
06	05	Mars in Conjunction with Sun
06	07:54	Pleiades 0.3°S of Moon
06	15	Mercury at Greatest Elong: 24.1°W
07	11:12	NEW MOON
08	10:43	Moon at Perigee: 358773 km
09	21:12	Pollux 2.2°N of Moon
10	17:35	Jupiter 3.4°S of Moon
12	07:35	Regulus 2.1°S of Moon
13	11	Neptune in Conjunction with Sun
14	01:53	FIRST QUARTER MOON
16	10:17	Spica 0.2°N of Moon
19	17:27	Mercury 3.6°N of Aldebaran
20	03:39	Antares 0.2°N of Moon
21	01:10	Summer Solstice
21	21:43	FULL MOON
23	08:23	Moon at Apogee: 406285 km
26	09	Saturn in Conjunction with Sun
29	23:54	LAST QUARTER MOON

Date	GMT (h:m)	Event	Date	GMT (h:m)	Event
Jul 03	18:11	Pleiades 0.5°S of Moon	Sep 30	00	Mercury at Greatest Elong: 17.9°W
05	23	Earth at Aphelion: 1.01669 AU	30	01:40	Mars 0.7°S of Moon: Occn.
06	17:53	NEW MOON			
06	19:35	Moon at Perigee: 357187 km	Oct 01	11:21	Mercury 1.4°N of Moon
08	08	Mercury at Superior Conjunction	02	18:49	NEW MOON
08	12:06	Jupiter 2.9°S of Moon	07	08:01	Antares 0.6°N of Moon
09	16:23	Regulus 1.8°S of Moon	10	10:27	FIRST QUARTER MOON
13	11:43	FIRST QUARTER MOON	10	21	Venus at Superior Conjunction
13	16:36	Spica 0.5°N of Moon	11	03:14	Moon at Apogee: 404271 km
17	09:29	Antares 0.3°N of Moon	18	08:18	FULL MOON
20	10:49	Moon at Apogee: 406330 km	20	21:13	Pleiades 0.7°S of Moon
21	12:47	FULL MOON	22	00	Orionid Meteor Shower
21	23	Venus 0.4°N of Saturn	24	12:04	Moon at Perigee: 369991 km
22	19	Mercury 1.0°N of Jupiter	24	16:06	Pollux 2.3°N of Moon
28	15	Delta-Aquarid Meteor Shower	25	06:28	LAST QUARTER MOON
29	10:04	LAST QUARTER MOON	27	04:42	Regulus 1.6°S of Moon
31	02:58	Pleiades 0.7°S of Moon	27	14:31	Jupiter 0.7°S of Moon: Occn.
31	04:57	Mercury 0.3°N of Regulus	28	15:57	Mars 1.3°N of Moon
			28	23	Mercury at Superior Conjunction
Aug 03	16:41	Venus 4.3°S of Moon			
03	18:08	Pollux 2.3°N of Moon	Nov 01	07:32	NEW MOON
03	19:31	Mars 3.5°S of Moon	03	16:44	Antares 0.4°N of Moon
04	04:41	Moon at Perigee: 358393 km	06	00	S Taurid Meteor Shower
05	00:40	NEW MOON	07	23:42	Moon at Apogee: 404510 km
06	02:46	Regulus 1.7°S of Moon	09	06:50	FIRST QUARTER MOON
06	15	Venus 0.6°S of Mars	10	02	Uranus in Conjunction with Sun
06	18:41	Mercury 1.4°S of Moon	13	00	N Taurid Meteor Shower
10	00:31	Spica 0.7°N of Moon	16	20:48	FULL MOON
12	00:21	FIRST QUARTER MOON	17	05:38	Pleiades 0.7°S of Moon
12	10	Jupiter in Conjunction with Sun	18	06	Leonid Meteor Shower
13	07	Perseid Meteor Shower	19	18:46	Moon at Perigee: 365580 km
13	15:53	Antares 0.5°N of Moon	20	22:09	Pollux 2.5°N of Moon
16	18:05	Moon at Apogee: 405741 km	23	10:11	Regulus 1.4°S of Moon
18	03	Mercury at Greatest Elong: 27.4°E	23	13:58	LAST QUARTER MOON
20	03:55	FULL MOON	24	02:24	Jupiter 0.0°N of Moon: Occn.
27	09:38	Pleiades 0.8°S of Moon	26	05:11	Mars 3.4°N of Moon
27	17:49	LAST QUARTER MOON	27	11:14	Spica 0.9°N of Moon
31	03:29	Pollux 2.2°N of Moon	28	21	Mars at Aphelion: 1.66613 AU
			30	23:01	NEW MOON
Sep 01	10:35	Moon at Perigee: 362122 km			
01	10:47	Mars 2.3°S of Moon	Dec 02	13:39	Mercury 2.5°N of Moon
02	04:39	Jupiter 1.9°S of Moon	05	19:32	Moon at Apogee: 405405 km
03	08:42	NEW MOON	09	03:28	FIRST QUARTER MOON
03	08:52	Partial Solar Eclipse; mag=0.975	12	10	Mercury at Greatest Elong: 20.7°E
04	12:22	Mercury 4.2°S of Moon	14	16:08	Pleiades 0.7°S of Moon
06	09:56	Spica 0.8°N of Moon	14	19	Geminid Meteor Shower
09	23:27	Antares 0.6°N of Moon	15	15	Neptune at Opposition
10	15:59	FIRST QUARTER MOON	16	08:17	FULL MOON
13	08:35	Moon at Apogee: 404840 km	17	19:22	Moon at Perigee: 360177 km
14	05	Mercury at Inferior Conjunction	18	06:44	Pollux 2.7°N of Moon
18	18:32	Total Lunar Eclipse; mag=1.150	20	06	Mercury 0.7°N of Venus
18	18:36	FULL MOON	20	16:53	Regulus 1.1°S of Moon
22	17:19	Autumnal Equinox	21	11:29	Jupiter 0.6°N of Moon: Occn.
23	15:04	Pleiades 0.8°S of Moon	21	14:42	Winter Solstice
26	00:11	LAST QUARTER MOON	22	23:40	LAST QUARTER MOON
27	02	Mars 0.5°N of Jupiter	23	03	Ursid Meteor Shower
27	04:03	Mars 0.7°N of Regulus	24	16:50	Spica 1.1°N of Moon
27	08:04	Jupiter 0.3°N of Regulus	27	04:26	Mars 3.4°N of Spica
27	10:36	Pollux 2.2°N of Moon	28	07:00	Antares 0.4°N of Moon
29	03:19	Moon at Perigee: 367253 km	29	23	Mercury at Inferior Conjunction
29	22:03	Regulus 1.7°S of Moon	30	16:57	NEW MOON
29	23:08	Jupiter 1.3°S of Moon			

Almanac of Astronomical Events for 2063

Date	GMT (h:m)	Event
Jan 01	12:27	Venus 2.1°N of Moon
02	10:03	Moon at Apogee: 406278 km
02	15	Saturn at Opposition
04	09	Quadrantid Meteor Shower
05	14	Earth at Perihelion: 0.98334 AU
07	22:16	FIRST QUARTER MOON
11	02:51	Pleiades 0.9°S of Moon
14	17:44	Pollux 2.8°N of Moon
14	19:11	FULL MOON
15	06:21	Moon at Perigee: 356937 km
17	02:19	Regulus 0.9°S of Moon
17	18:59	Jupiter 0.9°N of Moon: Occn.
20	15	Mercury at Greatest Elong: 24.2°W
20	23:12	Spica 1.4°N of Moon
21	12:05	LAST QUARTER MOON
24	12:42	Antares 0.6°N of Moon
29	12:23	NEW MOON
29	13:00	Moon at Apogee: 406600 km
Feb 01	01:47	Venus 1.2°S of Moon
06	13:37	FIRST QUARTER MOON
07	11:44	Pleiades 1.0°S of Moon
11	05:08	Pollux 2.7°N of Moon
12	19:32	Moon at Perigee: 356965 km
13	05:48	FULL MOON
13	13:41	Regulus 0.8°S of Moon
14	01:24	Jupiter 0.8°N of Moon: Occn.
17	07:49	Spica 1.5°N of Moon
20	03:07	LAST QUARTER MOON
20	19:06	Antares 0.7°N of Moon
23	20	Jupiter at Opposition
25	16:30	Moon at Apogee: 406312 km
28	07:38	NEW MOON
28	07:41	Annular Solar Eclipse; mag=0.929
Mar 03	11:23	Venus 3.7°S of Moon
06	18:10	Pleiades 1.1°S of Moon
08	01:06	FIRST QUARTER MOON
08	06	Mercury at Superior Conjunction
10	14:36	Pollux 2.7°N of Moon
13	00:43	Regulus 0.8°S of Moon
13	05:45	Moon at Perigee: 360132 km
13	06:46	Jupiter 0.4°N of Moon: Occn.
14	16:04	Partial Lunar Eclipse; mag=0.034
14	16:14	FULL MOON
16	18:13	Spica 1.5°N of Moon
20	03:08	Antares 0.7°N of Moon
20	13:59	Vernal Equinox
21	20:16	LAST QUARTER MOON
25	07:06	Moon at Apogee: 405402 km
30	00:50	NEW MOON
31	16:41	Mercury 1.7°S of Moon
Apr 02	11:19	Venus 3.7°S of Moon
02	16	Mercury at Greatest Elong: 19.0°E
02	23:34	Pleiades 1.0°S of Moon
06	09:18	FIRST QUARTER MOON

Date	GMT (h:m)	Event
Apr 06	21:24	Pollux 2.8°N of Moon
09	00:52	Venus 2.3°S of Pleiades
09	09:29	Regulus 0.8°S of Moon
09	11:36	Jupiter 0.2°N of Moon: Occn.
10	05:55	Moon at Perigee: 365244 km
13	02:34	FULL MOON
13	04:33	Spica 1.5°N of Moon
16	12:24	Antares 0.6°N of Moon
20	14:42	LAST QUARTER MOON
21	12	Mercury at Inferior Conjunction
22	02:01	Moon at Apogee: 404422 km
23	07	Lyrid Meteor Shower
27	20:35	Jupiter 0.5°N of Regulus
28	14:52	NEW MOON
30	05:54	Pleiades 0.9°S of Moon
May 02	03:01	Venus 2.2°S of Moon
04	02:50	Pollux 3.0°N of Moon
05	15:20	FIRST QUARTER MOON
05	20	Eta-Aquarid Meteor Shower
06	15:50	Regulus 0.6°S of Moon
06	17:27	Jupiter 0.3°N of Moon: Occn.
06	19:05	Moon at Perigee: 369614 km
10	13:11	Spica 1.5°N of Moon
10	20	Uranus at Opposition
12	13:11	FULL MOON
13	21:32	Antares 0.5°N of Moon
14	23	Mars at Opposition
17	20	Venus 3.3°N of Saturn
19	05	Mercury at Greatest Elong: 25.8°W
19	21:24	Moon at Apogee: 404086 km
20	09:16	LAST QUARTER MOON
23	17	Venus at Greatest Elong: 45.4°E
28	01:47	NEW MOON
29	03:02	Venus 3.8°S of Pollux
31	09:03	Pollux 3.2°N of Moon
31	11:13	Venus 0.7°S of Moon: Occn.
Jun 01	01:18	Moon at Perigee: 367758 km
02	21:22	Regulus 0.3°S of Moon
03	02:15	Jupiter 0.7°N of Moon: Occn.
03	20:28	FIRST QUARTER MOON
06	19:46	Spica 1.7°N of Moon
08	12:45	Mars 3.4°N of Moon
10	05:24	Antares 0.4°N of Moon
11	00:43	FULL MOON
16	01	Neptune in Conjunction with Sun
16	15:34	Moon at Apogee: 404594 km
19	02:43	LAST QUARTER MOON
21	07:02	Summer Solstice
22	19	Mercury at Superior Conjunction
23	23:36	Pleiades 1.0°S of Moon
26	10:25	NEW MOON
28	13:31	Moon at Perigee: 362850 km
29	03:08	Venus 1.8°S of Moon
30	04:03	Regulus 0.0°S of Moon
30	15:06	Jupiter 1.2°N of Moon: Occn.

91

Date	GMT (h:m)	Event
Jul 03	02:01	FIRST QUARTER MOON
04	01:19	Spica 2.0°N of Moon
04	15	Earth at Aphelion: 1.01668 AU
05	17:26	Mars 2.8°N of Moon
07	11:45	Antares 0.5°N of Moon
10	13:48	FULL MOON
11	12	Saturn in Conjunction with Sun
14	06:53	Moon at Apogee: 405517 km
16	15	Mercury 4.5°N of Venus
18	18:05	LAST QUARTER MOON
21	09:11	Pleiades 1.1°S of Moon
25	17:55	NEW MOON
26	01:01	Mercury 0.7°S of Regulus
26	17:32	Moon at Perigee: 358907 km
27	12:57	Regulus 0.1°N of Moon
27	15:07	Mercury 0.9°S of Moon: Occn.
28	07:55	Jupiter 1.9°N of Moon
28	22	Delta-Aquarid Meteor Shower
31	07:34	Spica 2.1°N of Moon
31	10	Mercury at Greatest Elong: 27.2°E
Aug 01	09:09	FIRST QUARTER MOON
01	19	Venus at Inferior Conjunction
02	14:09	Mars 2.8°N of Moon
03	17:19	Antares 0.7°N of Moon
07	11	Mercury 4.3°S of Jupiter
09	04:40	FULL MOON
10	16:02	Moon at Apogee: 406175 km
13	13	Perseid Meteor Shower
17	07:01	LAST QUARTER MOON
17	17:31	Pleiades 1.2°S of Moon
21	13:14	Saturn 3.7°S of Moon
21	13:59	Pollux 3.3°N of Moon
24	01:17	NEW MOON
24	01:20	Total Solar Eclipse; mag=1.075
24	02:53	Moon at Perigee: 357256 km
27	15:48	Spica 2.2°N of Moon
28	02	Mercury at Inferior Conjunction
30	19:04	FIRST QUARTER MOON
30	22:25	Mars 2.9°N of Moon
30	23:30	Antares 0.7°N of Moon
Sep 01	12:20	Mars 2.1°N of Antares
06	18:12	Moon at Apogee: 406190 km
07	20:39	Pen. Lunar Eclipse; mag=0.810
07	20:53	FULL MOON
10	04:01	Mercury 1.1°S of Regulus
13	11	Jupiter in Conjunction with Sun
13	15	Mercury at Greatest Elong: 17.9°W
14	00:01	Pleiades 1.2°S of Moon
15	17:44	LAST QUARTER MOON
17	23:18	Pollux 3.3°N of Moon
18	03:14	Saturn 3.4°S of Moon
20	10:18	Regulus 0.1°N of Moon
21	13:02	Moon at Perigee: 358430 km
22	09:21	NEW MOON
22	23:08	Autumnal Equinox
24	01:54	Spica 2.2°N of Moon
27	07:20	Antares 0.6°N of Moon
28	14:38	Mars 2.6°N of Moon
29	08:39	FIRST QUARTER MOON

Date	GMT (h:m)	Event
Oct 04	01:08	Moon at Apogee: 405611 km
07	13:27	FULL MOON
09	12:18	Venus 1.6°S of Regulus
10	03	Mercury at Superior Conjunction
11	05:29	Pleiades 1.1°S of Moon
12	01	Venus at Greatest Elong: 46.3°W
15	02:49	LAST QUARTER MOON
15	06:22	Pollux 3.5°N of Moon
15	13:40	Saturn 3.1°S of Moon
17	19:25	Regulus 0.2°N of Moon
18	10:32	Venus 0.4°N of Moon: Occn.
19	20:10	Moon at Perigee: 362294 km
19	20:36	Jupiter 3.7°N of Moon
21	18:46	NEW MOON
22	06	Orionid Meteor Shower
24	16:49	Antares 0.5°N of Moon
27	12:54	Mars 1.7°N of Moon
29	02:13	FIRST QUARTER MOON
31	16:27	Moon at Apogee: 404751 km
Nov 06	05:22	FULL MOON
07	10	Mars at Perihelion: 1.38112 AU
07	11:37	Pleiades 0.9°S of Moon
09	01	Venus 0.3°N of Jupiter
11	11:52	Pollux 3.8°N of Moon
11	20:20	Saturn 2.7°S of Moon
13	05:15	Mercury 2.2°N of Antares
13	10:56	LAST QUARTER MOON
14	02:08	Regulus 0.5°N of Moon
14	21	Uranus in Conjunction with Sun
16	13:06	Moon at Perigee: 367684 km
16	13:56	Jupiter 4.4°N of Moon
17	21:54	Spica 2.2°N of Moon
18	12	Leonid Meteor Shower
20	06:09	NEW MOON
21	23:10	Mercury 2.1°N of Moon
25	06	Mercury at Greatest Elong: 22.0°E
25	16:01	Mars 0.2°N of Moon: Occn.
27	22:59	FIRST QUARTER MOON
28	12:23	Moon at Apogee: 404245 km
Dec 04	19:37	Pleiades 0.9°S of Moon
05	20:06	FULL MOON
08	17:59	Pollux 4.0°N of Moon
09	01:03	Saturn 2.5°S of Moon
11	07:38	Regulus 0.8°N of Moon
11	13:44	Moon at Perigee: 370056 km
12	18:49	LAST QUARTER MOON
14	06	Mercury at Inferior Conjunction
15	01	Geminid Meteor Shower
15	04:51	Spica 2.4°N of Moon
18	02	Neptune at Opposition
18	11:24	Antares 0.4°N of Moon
19	20:04	NEW MOON
21	20:22	Winter Solstice
23	10	Ursid Meteor Shower
24	22:13	Mars 1.6°S of Moon
26	09:48	Moon at Apogee: 404532 km
27	20:57	FIRST QUARTER MOON

Almanac of Astronomical Events for 2064

Date	GMT (h:m)	Event
Jan 01	05:14	Pleiades 1.1°S of Moon
03	01	Mercury at Greatest Elong: 22.7°W
04	09:31	FULL MOON
04	12	Earth at Perihelion: 0.98329 AU
04	15	Quadrantid Meteor Shower
05	02:30	Pollux 4.1°N of Moon
05	06:22	Saturn 2.5°S of Moon
07	03:30	Moon at Perigee: 365135 km
07	14:25	Regulus 1.0°N of Moon
11	03:14	LAST QUARTER MOON
11	10:22	Spica 2.6°N of Moon
14	17:58	Antares 0.5°N of Moon
16	17	Saturn at Opposition
16	19:54	Mercury 3.8°N of Moon
18	12:37	NEW MOON
23	04:49	Mars 3.5°S of Moon
23	05:31	Moon at Apogee: 405424 km
26	17:42	FIRST QUARTER MOON
28	14:54	Pleiades 1.2°S of Moon
Feb 01	13:08	Pollux 4.0°N of Moon
01	13:22	Saturn 2.6°S of Moon
02	21:37	FULL MOON
02	21:47	Partial Lunar Eclipse; mag=0.038
03	23:54	Regulus 1.0°N of Moon
04	06:21	Moon at Perigee: 359865 km
07	16:48	Spica 2.7°N of Moon
09	12:55	LAST QUARTER MOON
10	23:25	Antares 0.5°N of Moon
14	23:52	Venus 2.0°N of Moon
17	06:58	Annular Solar Eclipse; mag=0.926
17	07:03	NEW MOON
18	18	Mercury at Superior Conjunction
19	18:41	Moon at Apogee: 406266 km
24	23:01	Pleiades 1.2°S of Moon
25	11:23	FIRST QUARTER MOON
28	21:24	Saturn 2.7°S of Moon
28	23:57	Pollux 4.0°N of Moon
Mar 02	11:10	Regulus 1.0°N of Moon
03	08:19	FULL MOON
03	17:31	Moon at Perigee: 356969 km
06	01:46	Spica 2.7°N of Moon
09	05:49	Antares 0.5°N of Moon
10	00:33	LAST QUARTER MOON
15	20	Mercury at Greatest Elong: 18.4°E
16	12:48	Venus 2.9°S of Moon
17	21:10	Moon at Apogee: 406581 km
18	01:45	NEW MOON
19	17:05	Mercury 1.4°S of Moon
19	19:40	Vernal Equinox
23	05:19	Pleiades 1.1°S of Moon
25	07	Jupiter at Opposition
26	01:13	FIRST QUARTER MOON
27	05:27	Saturn 2.6°S of Moon
27	08:57	Pollux 4.2°N of Moon
29	22:01	Regulus 1.1°N of Moon

Date	GMT (h:m)	Event
Apr 01	05:29	Moon at Perigee: 357235 km
01	17:40	FULL MOON
02	00	Mercury at Inferior Conjunction
02	12:37	Spica 2.6°N of Moon
05	14:24	Antares 0.3°N of Moon
08	14:25	LAST QUARTER MOON
14	02:09	Moon at Apogee: 406249 km
15	02:57	Mercury 4.4°S of Moon
16	19:02	NEW MOON
19	10:53	Pleiades 0.9°S of Moon
22	13	Lyrid Meteor Shower
23	13:34	Saturn 2.3°S of Moon
23	22:55	Mars 3.5°S of Pleiades
24	11:17	FIRST QUARTER MOON
26	06:41	Regulus 1.3°N of Moon
29	13:49	Moon at Perigee: 360386 km
29	23:23	Spica 2.6°N of Moon
30	00	Mercury at Greatest Elong: 27.0°W
May 01	02:08	FULL MOON
03	00:31	Antares 0.2°N of Moon
05	02	Eta-Aquarid Meteor Shower
08	06:16	LAST QUARTER MOON
11	16:10	Moon at Apogee: 405314 km
14	20	Uranus at Opposition
16	09:55	NEW MOON
18	00:53	Mars 3.7°S of Moon
20	22:44	Saturn 1.8°S of Moon
23	13:03	Regulus 1.6°N of Moon
23	18:15	FIRST QUARTER MOON
24	06	Venus at Superior Conjunction
27	08:19	Spica 2.8°N of Moon
27	12:05	Moon at Perigee: 365280 km
30	10:31	Antares 0.1°N of Moon
30	10:36	FULL MOON
Jun 06	07	Mercury at Superior Conjunction
06	23:23	LAST QUARTER MOON
08	09:54	Moon at Apogee: 404403 km
13	00:33	Pleiades 0.9°S of Moon
14	22:20	NEW MOON
17	10:01	Saturn 1.4°S of Moon
17	15	Neptune in Conjunction with Sun
19	18:34	Regulus 1.8°N of Moon
20	12:47	Summer Solstice
21	23:13	FIRST QUARTER MOON
23	00:41	Moon at Perigee: 369450 km
23	14:59	Spica 2.9°N of Moon
26	18:58	Antares 0.2°N of Moon
28	20:08	FULL MOON

Date	GMT (h:m)	Event
Jul 06	04:32	Moon at Apogee: 404174 km
06	10	Earth at Aphelion: 1.01670 AU
06	16:48	LAST QUARTER MOON
10	09:05	Pleiades 1.0°S of Moon
12	11	Mercury at Greatest Elong: 26.4°E
14	08:46	NEW MOON
16	08:01	Mercury 0.9°S of Moon: Occn.
17	01:07	Regulus 1.9°N of Moon
18	08:38	Moon at Perigee: 367671 km
20	09	Mars in Conjunction with Sun
20	20:27	Spica 3.1°N of Moon
21	03:35	FIRST QUARTER MOON
24	01:26	Antares 0.2°N of Moon
25	12	Saturn in Conjunction with Sun
28	04	Delta-Aquarid Meteor Shower
28	07:40	FULL MOON
28	07:51	Partial Lunar Eclipse; mag=0.104
Aug 02	06:07	Venus 1.0°N of Regulus
02	22:45	Moon at Apogee: 404778 km
05	09:41	LAST QUARTER MOON
06	17:47	Pleiades 1.1°S of Moon
08	22	Jupiter at Aphelion: 5.45640 AU
09	08	Mercury at Inferior Conjunction
12	17:44	Total Solar Eclipse; mag=1.049
12	17:49	NEW MOON
12	20	Perseid Meteor Shower
14	11:54	Venus 4.4°N of Moon
14	20:26	Moon at Perigee: 362724 km
17	02:39	Spica 3.1°N of Moon
19	08:55	FIRST QUARTER MOON
20	06:52	Antares 0.2°N of Moon
26	21:35	FULL MOON
27	02	Mercury at Greatest Elong: 18.3°W
30	14:46	Moon at Apogee: 405759 km
Sep 03	01:36	Pleiades 1.0°S of Moon
04	01:29	LAST QUARTER MOON
06	14	Venus 0.4°S of Jupiter
07	07:55	Mars 0.7°N of Regulus
08	07:29	Saturn 0.2°S of Moon: Occn.
09	20:11	Regulus 1.9°N of Moon
09	23:29	Mars 2.9°N of Moon
11	02:11	NEW MOON
12	00:51	Moon at Perigee: 358598 km
13	11:01	Spica 3.0°N of Moon
15	13:59	Venus 2.1°N of Spica
16	13:01	Antares 0.1°N of Moon
17	16:45	FIRST QUARTER MOON
21	07	Mercury at Superior Conjunction
22	04:58	Autumnal Equinox
25	13:38	FULL MOON
27	00:14	Moon at Apogee: 406395 km
30	08:06	Pleiades 0.8°S of Moon

Date	GMT (h:m)	Event
Oct 03	15:50	LAST QUARTER MOON
05	22:00	Saturn 0.2°N of Moon: Occn.
07	06:47	Regulus 2.1°N of Moon
08	16:54	Mars 4.7°N of Moon
10	10:34	NEW MOON
10	11:14	Moon at Perigee: 356866 km
13	02:06	Venus 3.8°N of Moon
13	06	Jupiter in Conjunction with Sun
13	21:20	Antares 0.1°S of Moon
15	19	Mars at Aphelion: 1.66617 AU
17	04:22	FIRST QUARTER MOON
21	12	Orionid Meteor Shower
23	16:41	Venus 2.8°N of Antares
24	01:15	Moon at Apogee: 406377 km
25	07:06	FULL MOON
27	13:51	Pleiades 0.7°S of Moon
Nov 02	04:24	LAST QUARTER MOON
02	08:58	Saturn 0.7°N of Moon: Occn.
03	15:47	Regulus 2.3°N of Moon
06	21	Mercury at Greatest Elong: 23.3°E
07	08:39	Spica 3.0°N of Moon
07	22:41	Moon at Perigee: 358227 km
08	06:50	Mercury 1.8°N of Antares
08	19:45	NEW MOON
10	07:46	Antares 0.2°S of Moon
10	11:09	Mercury 1.6°N of Moon
11	21:51	Venus 0.7°N of Moon: Occn.
15	20:14	FIRST QUARTER MOON
17	18	Leonid Meteor Shower
18	16	Uranus in Conjunction with Sun
20	08:27	Moon at Apogee: 405800 km
23	20:00	Pleiades 0.6°S of Moon
24	00:58	FULL MOON
27	14	Mercury at Inferior Conjunction
29	15:39	Saturn 1.1°N of Moon: Occn.
30	22:23	Regulus 2.6°N of Moon
Dec 01	15:00	LAST QUARTER MOON
02	05:49	Mars 2.9°N of Spica
04	18:15	Spica 3.1°N of Moon
06	06:12	Moon at Perigee: 362504 km
08	06:28	NEW MOON
11	21:49	Venus 2.5°S of Moon
14	07	Geminid Meteor Shower
15	15:45	FIRST QUARTER MOON
15	17	Mercury at Greatest Elong: 21.3°W
18	01:06	Moon at Apogee: 404936 km
18	21	Mars 0.1°S of Jupiter
19	13	Neptune at Opposition
21	02:10	Winter Solstice
21	03:23	Pleiades 0.7°S of Moon
22	16	Ursid Meteor Shower
23	18:14	FULL MOON
26	19:37	Saturn 1.3°N of Moon
28	03:53	Regulus 2.8°N of Moon
29	05	Venus at Greatest Elong: 47.3°E
30	23:50	LAST QUARTER MOON

Almanac of Astronomical Events for 2065

Date	GMT (h:m)	Event
Jan 01	01:09	Spica 3.3°N of Moon
02	16	Earth at Perihelion: 0.98334 AU
02	20:34	Moon at Perigee: 368142 km
03	22	Quadrantid Meteor Shower
04	03:49	Antares 0.2°S of Moon
06	19:15	NEW MOON
10	18:00	Venus 3.2°S of Moon
14	13:19	FIRST QUARTER MOON
14	21:51	Moon at Apogee: 404429 km
17	11:50	Pleiades 0.8°S of Moon
22	09:53	FULL MOON
22	09:57	Total Lunar Eclipse; mag=1.223
22	23:38	Saturn 1.2°N of Moon: Occn.
24	10:37	Regulus 2.8°N of Moon
27	15:26	Moon at Perigee: 369668 km
28	06:33	Spica 3.4°N of Moon
29	07:38	LAST QUARTER MOON
29	16	Saturn at Opposition
30	09	Mercury at Superior Conjunction
31	10:23	Antares 0.2°S of Moon
Feb 05	09:50	Partial Solar Eclipse; mag=0.912
05	10:02	NEW MOON
08	14:32	Venus 0.4°N of Moon: Occn.
11	18:57	Moon at Apogee: 404706 km
13	10:50	FIRST QUARTER MOON
13	20:23	Pleiades 0.7°S of Moon
19	05:24	Saturn 1.0°N of Moon: Occn.
20	19:44	Regulus 2.8°N of Moon
20	23:11	FULL MOON
23	13:24	Moon at Perigee: 364457 km
24	13:00	Spica 3.3°N of Moon
27	06	Mercury at Greatest Elong: 18.1°E
27	15:29	LAST QUARTER MOON
27	15:43	Antares 0.2°S of Moon
28	04:54	Mars 4.4°N of Moon
Mar 07	02:15	NEW MOON
11	03	Venus at Inferior Conjunction
11	13:11	Moon at Apogee: 405573 km
13	04:05	Pleiades 0.6°S of Moon
15	06:25	FIRST QUARTER MOON
15	07	Mercury at Inferior Conjunction
18	12:56	Saturn 1.0°N of Moon: Occn.
20	01:27	Vernal Equinox
20	06:21	Regulus 2.9°N of Moon
22	09:56	FULL MOON
23	16:36	Moon at Perigee: 359485 km
23	22:00	Spica 3.2°N of Moon
26	22:15	Antares 0.4°S of Moon
28	12:55	Mars 2.6°N of Moon
29	00:24	LAST QUARTER MOON

Date	GMT (h:m)	Event
Apr 02	23:08	Venus 2.4°N of Moon
05	19:01	NEW MOON
08	00:30	Moon at Apogee: 406389 km
09	10:41	Pleiades 0.4°S of Moon
12	01	Mercury at Greatest Elong: 27.7°W
13	22:38	FIRST QUARTER MOON
14	21:24	Saturn 1.2°N of Moon
16	16:33	Regulus 3.1°N of Moon
20	08:50	Spica 3.1°N of Moon
20	18:36	FULL MOON
21	02:34	Moon at Perigee: 356951 km
22	19	Lyrid Meteor Shower
23	07:14	Antares 0.6°S of Moon
25	00	Jupiter at Opposition
25	19:33	Mars 0.4°N of Moon: Occn.
27	11:02	LAST QUARTER MOON
May 01	13:49	Venus 4.3°S of Moon
05	03:19	Moon at Apogee: 406674 km
05	08	Eta-Aquarid Meteor Shower
05	11:30	NEW MOON
12	06:28	Saturn 1.6°N of Moon
13	10:52	FIRST QUARTER MOON
14	00:50	Regulus 3.3°N of Moon
17	19:33	Spica 3.2°N of Moon
19	12:57	Moon at Perigee: 357407 km
19	19	Uranus at Opposition
20	02:05	FULL MOON
20	04	Venus at Greatest Elong: 46.0°W
20	17:52	Antares 0.7°S of Moon
21	18	Mercury at Superior Conjunction
23	21:15	Mars 2.1°S of Moon
26	23:38	LAST QUARTER MOON
Jun 01	09:08	Moon at Apogee: 406272 km
04	03:05	NEW MOON
05	14:54	Mercury 0.2°S of Moon: Occn.
08	16:27	Saturn 2.0°N of Moon
10	07:06	Regulus 3.6°N of Moon
11	19:25	FIRST QUARTER MOON
14	04:28	Spica 3.4°N of Moon
16	20:01	Moon at Perigee: 360552 km
17	04:22	Antares 0.7°S of Moon
18	09:28	FULL MOON
20	05	Neptune in Conjunction with Sun
20	11:26	Mars 4.4°S of Moon
20	18:31	Summer Solstice
24	05	Mercury at Greatest Elong: 25.2°E
25	14:08	LAST QUARTER MOON
28	22:35	Moon at Apogee: 405316 km
30	05:26	Pleiades 0.3°S of Moon

Date	GMT (h:m)	Event
Jul 03	17:16	NEW MOON
03	17:32	Partial Solar Eclipse; mag=0.164
05	02	Earth at Aphelion: 1.01671 AU
05	12:23	Mercury 1.3°S of Moon
06	03:58	Saturn 2.4°N of Moon
07	12:36	Regulus 3.7°N of Moon
11	01:16	FIRST QUARTER MOON
11	06:25	Venus 3.3°N of Aldebaran
11	11:06	Spica 3.5°N of Moon
13	20	Mars at Opposition
14	13:08	Antares 0.7°S of Moon
14	17:50	Moon at Perigee: 365373 km
17	17:45	FULL MOON
17	17:47	Total Lunar Eclipse; mag=1.612
21	19	Mercury at Inferior Conjunction
25	06:22	LAST QUARTER MOON
26	15:50	Moon at Apogee: 404434 km
27	12:59	Pleiades 0.4°S of Moon
28	10	Delta-Aquarid Meteor Shower
30	03:46	Venus 3.6°S of Moon
Aug 02	05:32	Partial Solar Eclipse; mag=0.490
02	05:46	NEW MOON
03	18:58	Regulus 3.6°N of Moon
07	16:28	Spica 3.4°N of Moon
09	05:52	FIRST QUARTER MOON
09	06	Saturn in Conjunction with Sun
10	04	Mercury at Greatest Elong: 19.0°W
10	06:04	Moon at Perigee: 369504 km
10	19:41	Antares 0.7°S of Moon
13	02	Perseid Meteor Shower
13	10:22	Mars 4.9°S of Moon
16	03:45	FULL MOON
23	10:35	Moon at Apogee: 404249 km
23	20:57	Pleiades 0.3°S of Moon
23	23:56	LAST QUARTER MOON
29	09:25	Venus 1.3°N of Moon
30	08:34	Saturn 3.1°N of Moon
31	16:39	NEW MOON
Sep 03	22:32	Spica 3.3°N of Moon
04	07	Mercury at Superior Conjunction
04	13:22	Moon at Perigee: 367552 km
07	01:02	Antares 0.9°S of Moon
07	10:49	FIRST QUARTER MOON
09	09	Venus 0.2°S of Saturn
09	23:56	Mars 4.1°S of Moon
14	16:05	FULL MOON
16	18:05	Venus 0.5°N of Regulus
20	04:42	Pleiades 0.1°S of Moon
20	05:27	Moon at Apogee: 404888 km
22	10:41	Autumnal Equinox
22	18:09	LAST QUARTER MOON
24	11	Mars at Perihelion: 1.38131 AU
27	00:21	Saturn 3.6°N of Moon
27	12:57	Regulus 3.7°N of Moon
28	17:04	Mercury 1.2°N of Spica
30	02:24	NEW MOON

Date	GMT (h:m)	Event
Oct 01	06:52	Spica 3.2°N of Moon
01	14:28	Mercury 4.2°N of Moon
02	02:38	Moon at Perigee: 362386 km
04	07:14	Antares 1.1°S of Moon
06	17:37	FIRST QUARTER MOON
08	03:26	Mars 4.2°S of Moon
13	06	Mercury 3.3°S of Jupiter
14	07:04	FULL MOON
17	11:44	Pleiades 0.1°N of Moon
17	22:01	Moon at Apogee: 405850 km
20	10	Mercury at Greatest Elong: 24.6°E
21	18	Orionid Meteor Shower
22	11:53	LAST QUARTER MOON
24	14:54	Saturn 4.2°N of Moon
24	22:55	Regulus 3.9°N of Moon
29	11:48	NEW MOON
30	08:37	Moon at Perigee: 358206 km
30	22:58	Mercury 1.2°N of Moon: Occn.
31	15:55	Antares 1.3°S of Moon
Nov 05	03:26	FIRST QUARTER MOON
05	19	S Taurid Meteor Shower
11	20	Mercury at Inferior Conjunction
12	15	Jupiter in Conjunction with Sun
12	18	N Taurid Meteor Shower
13	00:37	FULL MOON
13	18:06	Pleiades 0.2°N of Moon
14	06:34	Moon at Apogee: 406402 km
18	00	Leonid Meteor Shower
21	02:04	Saturn 4.7°N of Moon
21	03:51	LAST QUARTER MOON
21	07:25	Regulus 4.2°N of Moon
23	09	Uranus in Conjunction with Sun
25	04:27	Spica 3.3°N of Moon
27	20:39	Moon at Perigee: 356718 km
27	21:40	NEW MOON
28	16	Mercury at Greatest Elong: 20.1°W
Dec 04	16:54	FIRST QUARTER MOON
06	02	Mercury 0.9°N of Jupiter
06	21:43	Saturn 1.6°N of Regulus
11	00:17	Pleiades 0.2°N of Moon
11	06:28	Moon at Apogee: 406338 km
12	19:52	FULL MOON
14	14	Geminid Meteor Shower
20	17:12	LAST QUARTER MOON
21	07:59	Winter Solstice
21	23	Neptune at Opposition
22	13:48	Spica 3.4°N of Moon
22	22	Ursid Meteor Shower
24	17	Venus at Superior Conjunction
25	00:31	Jupiter 4.8°N of Moon
25	14:13	Antares 1.3°S of Moon
26	09:01	Moon at Perigee: 358544 km
27	08:27	NEW MOON
27	08:38	Partial Solar Eclipse; mag=0.877

Almanac of Astronomical Events for 2066

Date	GMT (h:m)	Event
Jan 03	09:56	FIRST QUARTER MOON
04	04	Quadrantid Meteor Shower
05	06	Earth at Perihelion: 0.98332 AU
07	06:53	Pleiades 0.1°N of Moon
07	15:46	Moon at Apogee: 405716 km
10	22	Mercury at Superior Conjunction
11	15:03	Total Lunar Eclipse; mag=1.138
11	15:07	FULL MOON
18	20:26	Spica 3.4°N of Moon
19	03:48	LAST QUARTER MOON
21	17:24	Jupiter 4.5°N of Moon
21	23:27	Antares 1.3°S of Moon
23	15:57	Moon at Perigee: 363177 km
25	20:14	NEW MOON
Feb 02	05:44	FIRST QUARTER MOON
03	14:18	Pleiades 0.2°N of Moon
04	09:58	Moon at Apogee: 404789 km
10	08:29	FULL MOON
10	19	Mercury at Greatest Elong: 18.2°E
12	11	Saturn at Opposition
15	01:44	Spica 3.3°N of Moon
17	12:14	LAST QUARTER MOON
18	05:50	Jupiter 4.1°N of Moon
18	05:53	Antares 1.4°S of Moon
20	01:15	Moon at Perigee: 368665 km
24	08:50	NEW MOON
26	07	Mercury at Inferior Conjunction
Mar 01	05:10	Mars 4.4°S of Moon
02	22:22	Pleiades 0.4°N of Moon
04	02:48	FIRST QUARTER MOON
04	06:48	Moon at Apogee: 404251 km
09	19:28	Saturn 4.7°N of Moon
11	22:48	FULL MOON
14	08:09	Spica 3.1°N of Moon
16	19:50	Moon at Perigee: 369243 km
17	11:10	Antares 1.7°S of Moon
17	14:13	Jupiter 3.8°N of Moon
18	19:25	LAST QUARTER MOON
20	07:19	Vernal Equinox
25	08	Mercury at Greatest Elong: 27.8°W
25	22:13	NEW MOON
30	02:47	Mars 2.9°S of Moon
30	06:28	Pleiades 0.6°N of Moon

Date	GMT (h:m)	Event
Apr 01	02:53	Moon at Apogee: 404532 km
02	23:09	FIRST QUARTER MOON
03	02:20	Mars 3.2°S of Pleiades
10	10:03	FULL MOON
10	16:53	Spica 3.1°N of Moon
12	22:32	Moon at Perigee: 364162 km
13	17:48	Antares 1.9°S of Moon
13	20:13	Jupiter 3.6°N of Moon
17	02:23	LAST QUARTER MOON
21	13:57	Venus 3.3°S of Pleiades
23	01	Lyrid Meteor Shower
24	12:29	NEW MOON
26	13:52	Pleiades 0.8°N of Moon
27	04:45	Venus 1.9°S of Moon
27	23:50	Mars 1.1°S of Moon: Occn.
28	19:52	Moon at Apogee: 405408 km
May 02	16:57	FIRST QUARTER MOON
05	14	Eta-Aquarid Meteor Shower
06	02	Mercury at Superior Conjunction
08	03:12	Spica 3.1°N of Moon
09	18:58	FULL MOON
11	01:15	Moon at Perigee: 359581 km
11	01:30	Jupiter 3.7°N of Moon
11	02:52	Antares 2.0°S of Moon
13	23	Venus 0.6°N of Mars
16	10:01	LAST QUARTER MOON
24	03:38	NEW MOON
24	17	Uranus at Opposition
25	23:57	Mercury 1.0°N of Moon: Occn.
26	06:49	Moon at Apogee: 406227 km
26	20:38	Mars 0.8°N of Moon: Occn.
26	21	Jupiter at Opposition
27	12:33	Venus 2.4°N of Moon
Jun 01	07:13	FIRST QUARTER MOON
04	13:24	Spica 3.2°N of Moon
05	19	Mercury at Greatest Elong: 23.6°E
07	07:00	Jupiter 3.9°N of Moon
07	13:32	Antares 2.0°S of Moon
08	02:31	FULL MOON
08	10:06	Moon at Perigee: 357249 km
10	06	Mercury 1.9°S of Mars
14	19:10	LAST QUARTER MOON
20	02:17	Pleiades 0.8°N of Moon
21	00:16	Summer Solstice
22	10:28	Moon at Apogee: 406486 km
22	19:15	NEW MOON
22	19:24	Annular Solar Eclipse; mag=0.943
22	20	Neptune in Conjunction with Sun
24	17:16	Mars 2.6°N of Moon
30	17:59	FIRST QUARTER MOON

Date	GMT (h:m)	Event
Jul 01	21:57	Spica 3.3°N of Moon
02	10	Mercury at Inferior Conjunction
02	21	Venus 0.3°N of Saturn
04	13:09	Jupiter 4.0°N of Moon
05	00:04	Antares 2.0°S of Moon
05	03	Earth at Aphelion: 1.01665 AU
06	20:02	Moon at Perigee: 357676 km
07	09:28	Partial Lunar Eclipse; mag=0.775
07	09:34	FULL MOON
07	20:53	Venus 0.9°N of Regulus
14	06:38	LAST QUARTER MOON
17	08:22	Pleiades 0.8°N of Moon
19	16:13	Moon at Apogee: 406051 km
20	15:27	Mercury 3.1°S of Moon
22	10:34	NEW MOON
23	21	Mercury at Greatest Elong: 20.1°W
28	16	Delta-Aquarid Meteor Shower
29	04:22	Spica 3.2°N of Moon
30	02:01	FIRST QUARTER MOON
31	20:10	Jupiter 3.9°N of Moon
Aug 01	08:50	Antares 2.1°S of Moon
03	03	Venus at Greatest Elong: 45.7°E
04	03:18	Moon at Perigee: 360710 km
05	16:59	FULL MOON
12	20:59	LAST QUARTER MOON
13	08	Perseid Meteor Shower
13	15:16	Pleiades 0.9°N of Moon
16	05:19	Moon at Apogee: 405122 km
18	20	Mercury at Superior Conjunction
21	00:50	NEW MOON
23	17	Saturn in Conjunction with Sun
24	17:40	Venus 1.8°N of Moon
25	09:40	Spica 3.0°N of Moon
27	10	Mars in Conjunction with Sun
28	04:35	Jupiter 3.5°N of Moon
28	08:25	FIRST QUARTER MOON
28	15:19	Antares 2.2°S of Moon
Sep 01	01:49	Moon at Perigee: 365541 km
02	22	Mars at Aphelion: 1.66617 AU
04	01:37	FULL MOON
05	23:40	Venus 2.5°S of Spica
09	23:10	Pleiades 1.1°N of Moon
11	14:16	LAST QUARTER MOON
12	22:49	Moon at Apogee: 404308 km
19	13:47	NEW MOON
21	13:35	Mercury 3.5°N of Moon
21	15:40	Spica 2.9°N of Moon
21	17:02	Venus 2.9°S of Moon
22	16:12	Mercury 0.3°N of Spica
22	16:27	Autumnal Equinox
24	15:22	Jupiter 2.9°N of Moon
24	20:36	Antares 2.5°S of Moon
26	14:19	FIRST QUARTER MOON
27	12:48	Moon at Perigee: 369827 km

Date	GMT (h:m)	Event
Oct 02	22	Mercury at Greatest Elong: 25.8°E
03	12:25	FULL MOON
07	07:37	Pleiades 1.3°N of Moon
10	18:32	Moon at Apogee: 404214 km
11	09:43	LAST QUARTER MOON
11	23	Venus at Inferior Conjunction
19	01:42	NEW MOON
22	00	Orionid Meteor Shower
22	02:51	Antares 2.7°S of Moon
22	05:31	Jupiter 2.3°N of Moon
22	18:37	Moon at Perigee: 367509 km
25	20:52	FIRST QUARTER MOON
26	23	Mercury at Inferior Conjunction
Nov 02	02:13	FULL MOON
03	15:46	Pleiades 1.5°N of Moon
06	01	S Taurid Meteor Shower
07	14:36	Moon at Apogee: 404915 km
10	05:45	LAST QUARTER MOON
11	22	Mercury at Greatest Elong: 19.1°W
12	07:21	Mars 2.7°N of Spica
13	00	N Taurid Meteor Shower
15	09:51	Spica 2.9°N of Moon
17	13:06	NEW MOON
18	06	Leonid Meteor Shower
18	23:25	Jupiter 1.6°N of Moon
19	10:44	Moon at Perigee: 362077 km
24	05:10	FIRST QUARTER MOON
28	02	Uranus in Conjunction with Sun
29	17:49	Venus 3.6°N of Spica
30	22:52	Pleiades 1.5°N of Moon
Dec 01	19:16	FULL MOON
05	07:41	Moon at Apogee: 405855 km
10	00:38	LAST QUARTER MOON
12	20:13	Spica 2.9°N of Moon
13	13	Jupiter in Conjunction with Sun
14	08:40	Mars 4.5°N of Moon
14	20	Geminid Meteor Shower
15	22:41	Antares 2.8°S of Moon
17	00:17	NEW MOON
17	00:22	Total Solar Eclipse; mag=1.042
17	18:55	Moon at Perigee: 357950 km
21	10	Mercury at Superior Conjunction
21	13:45	Winter Solstice
22	17	Venus at Greatest Elong: 46.9°W
23	04	Ursid Meteor Shower
23	16:07	FIRST QUARTER MOON
24	10	Neptune at Opposition
28	04:55	Pleiades 1.5°N of Moon
31	14:28	Pen. Lunar Eclipse; mag=0.977
31	14:41	FULL MOON

Almanac of Astronomical Events for 2067

Date	GMT (h:m)	Event
Jan 01	14:49	Moon at Apogee: 406354 km
03	17	Earth at Perihelion: 0.98326 AU
04	10	Quadrantid Meteor Shower
07	02	Venus 2.7°N of Mars
08	17:01	LAST QUARTER MOON
09	04:50	Spica 2.9°N of Moon
12	01:35	Mars 2.6°N of Moon
12	09:42	Antares 2.8°S of Moon
13	17:28	Jupiter 0.4°N of Moon: Occn.
15	07:56	Moon at Perigee: 356756 km
15	11:17	NEW MOON
22	06:17	FIRST QUARTER MOON
24	10:50	Pleiades 1.5°N of Moon
25	07	Mercury at Greatest Elong: 18.6°E
28	14:45	Moon at Apogee: 406278 km
30	10:29	FULL MOON
Feb 04	21	Venus 1.6°N of Jupiter
05	11:07	Spica 2.7°N of Moon
07	06:14	LAST QUARTER MOON
08	18:33	Antares 2.9°S of Moon
09	18:02	Mars 0.5°N of Moon: Occn.
09	19	Mercury at Inferior Conjunction
10	12:52	Jupiter 0.3°S of Moon: Occn.
10	22:51	Venus 0.5°N of Moon: Occn.
12	19:54	Moon at Perigee: 358905 km
13	21:57	NEW MOON
20	17:51	Pleiades 1.7°N of Moon
20	23:30	FIRST QUARTER MOON
25	01:39	Moon at Apogee: 405610 km
26	02	Saturn at Opposition
Mar 01	04:42	FULL MOON
04	01	Mars 0.6°S of Jupiter
04	16:28	Spica 2.5°N of Moon
07	17	Mercury at Greatest Elong: 27.3°W
08	00:45	Antares 3.2°S of Moon
08	16:16	LAST QUARTER MOON
10	04:17	Jupiter 0.9°S of Moon: Occn.
10	10:13	Mars 1.9°S of Moon
12	20:08	Venus 4.4°S of Moon
13	00:40	Moon at Perigee: 363617 km
15	08:29	NEW MOON
20	02:22	Pleiades 2.0°N of Moon
20	12:55	Vernal Equinox
22	18:44	FIRST QUARTER MOON
24	19:49	Moon at Apogee: 404668 km
30	20:08	FULL MOON
31	22:50	Spica 2.4°N of Moon

Date	GMT (h:m)	Event
Apr 04	06:04	Antares 3.4°S of Moon
06	15:04	Jupiter 1.5°S of Moon
06	23:37	LAST QUARTER MOON
08	02:23	Mars 4.4°S of Moon
09	05:49	Moon at Perigee: 368765 km
13	19:23	NEW MOON
16	11:31	Pleiades 2.2°N of Moon
20	06	Mercury at Superior Conjunction
21	14:15	FIRST QUARTER MOON
21	15:37	Moon at Apogee: 404190 km
23	07	Lyrid Meteor Shower
28	07:02	Spica 2.4°N of Moon
29	08:40	FULL MOON
May 01	12:42	Antares 3.5°S of Moon
03	22:03	Jupiter 1.8°S of Moon
04	04:17	Moon at Perigee: 368880 km
04	18:07	Mercury 2.2°S of Pleiades
05	20	Eta-Aquarid Meteor Shower
06	05:19	LAST QUARTER MOON
13	07:20	NEW MOON
15	03:01	Mercury 2.0°N of Moon
18	12	Mercury at Greatest Elong: 22.1°E
19	10:35	Moon at Apogee: 404564 km
21	08:29	FIRST QUARTER MOON
25	16:27	Spica 2.5°N of Moon
28	18:42	FULL MOON
28	18:54	Pen. Lunar Eclipse; mag=0.640
28	21:29	Antares 3.6°S of Moon
29	15	Uranus at Opposition
31	02:57	Jupiter 1.8°S of Moon
31	07:21	Moon at Perigee: 364053 km
Jun 04	10:38	LAST QUARTER MOON
10	02:55	Pleiades 2.2°N of Moon
11	20:40	Annular Solar Eclipse; mag=0.967
11	20:41	NEW MOON
12	08	Mercury at Inferior Conjunction
16	03:04	Moon at Apogee: 405510 km
20	00:28	FIRST QUARTER MOON
21	05:56	Summer Solstice
22	01:44	Spica 2.5°N of Moon
25	07:41	Antares 3.6°S of Moon
25	10	Neptune in Conjunction with Sun
27	02:39	Pen. Lunar Eclipse; mag=0.375
27	02:52	FULL MOON
27	07:27	Jupiter 1.5°S of Moon
28	09:11	Moon at Perigee: 359568 km
29	14	Jupiter at Opposition

Date	GMT (h:m)	Event
Jul 03	17:02	LAST QUARTER MOON
06	02	Mercury at Greatest Elong: 21.5°W
07	04	Earth at Aphelion: 1.01673 AU
07	08:40	Pleiades 2.2°N of Moon
09	13:48	Mercury 1.8°S of Moon
11	11:16	NEW MOON
13	14:29	Moon at Apogee: 406361 km
19	09:39	Spica 2.4°N of Moon
19	13:59	FIRST QUARTER MOON
22	17:43	Antares 3.6°S of Moon
24	12:46	Jupiter 1.3°S of Moon
26	09:58	FULL MOON
26	17:23	Moon at Perigee: 357147 km
28	22	Delta-Aquarid Meteor Shower
31	19	Venus at Superior Conjunction
Aug 02	01:51	LAST QUARTER MOON
02	19	Mercury at Superior Conjunction
03	14:18	Pleiades 2.4°N of Moon
09	18:29	Moon at Apogee: 406600 km
10	02:36	NEW MOON
12	07	Mars at Perihelion: 1.38133 AU
13	14	Perseid Meteor Shower
15	15:52	Spica 2.2°N of Moon
18	01:09	FIRST QUARTER MOON
19	01	Mercury 0.7°S of Saturn
19	02:08	Antares 3.8°S of Moon
20	19:28	Jupiter 1.4°S of Moon
24	03:23	Moon at Perigee: 357474 km
24	16:57	FULL MOON
30	21:06	Pleiades 2.6°N of Moon
31	14:04	LAST QUARTER MOON
Sep 05	23:25	Moon at Apogee: 406157 km
06	19	Saturn in Conjunction with Sun
08	18:09	NEW MOON
11	04:21	Mercury 2.5°N of Moon
11	21:16	Spica 2.0°N of Moon
15	08:27	Antares 4.1°S of Moon
15	10	Mercury at Greatest Elong: 26.7°E
16	10:20	FIRST QUARTER MOON
17	03:45	Jupiter 1.7°S of Moon
20	16:19	Mercury 1.0°S of Spica
21	11:31	Moon at Perigee: 360582 km
22	22:20	Autumnal Equinox
23	00:54	FULL MOON
27	05:39	Pleiades 2.9°N of Moon
30	06:01	LAST QUARTER MOON
30	13:50	Venus 2.6°N of Spica

Date	GMT (h:m)	Event
Oct 02	05	Mercury 4.3°S of Venus
02	18	Mars at Opposition
03	12:21	Moon at Apogee: 405262 km
08	09:28	NEW MOON
10	02:39	Venus 3.6°N of Moon
10	21	Mercury at Inferior Conjunction
12	13:46	Antares 4.3°S of Moon
14	14:05	Jupiter 2.2°S of Moon
15	18:03	FIRST QUARTER MOON
19	10:41	Moon at Perigee: 365700 km
22	06	Orionid Meteor Shower
22	10:56	FULL MOON
24	15:21	Pleiades 3.0°N of Moon
26	11	Mercury at Greatest Elong: 18.4°W
30	01:08	LAST QUARTER MOON
31	06:44	Moon at Apogee: 404504 km
Nov 05	11:01	Spica 1.9°N of Moon
06	07	S Taurid Meteor Shower
06	16:26	Venus 3.7°N of Antares
07	00:14	NEW MOON
09	02:34	Venus 1.0°S of Moon: Occn.
11	03:21	Jupiter 2.7°S of Moon
13	06	N Taurid Meteor Shower
14	01:07	FIRST QUARTER MOON
14	16:11	Moon at Perigee: 370093 km
18	13	Leonid Meteor Shower
20	23:49	FULL MOON
21	00:03	Pen. Lunar Eclipse; mag=0.654
21	00:43	Pleiades 3.1°N of Moon
28	03:30	Moon at Apogee: 404463 km
28	22:06	LAST QUARTER MOON
30	12	Mercury at Superior Conjunction
Dec 02	17	Uranus in Conjunction with Sun
02	20:08	Spica 1.9°N of Moon
05	11	Venus 1.6°S of Jupiter
06	14:02	Hybrid Solar Eclipse; mag=1.001
06	14:05	NEW MOON
08	20:16	Jupiter 3.3°S of Moon
10	00:36	Moon at Perigee: 367040 km
13	08:38	FIRST QUARTER MOON
15	02	Geminid Meteor Shower
15	03:21	Mars 4.5°S of Moon
18	08:21	Pleiades 3.0°N of Moon
20	15:41	FULL MOON
21	19:44	Winter Solstice
23	10	Ursid Meteor Shower
25	23:56	Moon at Apogee: 405164 km
26	20	Neptune at Opposition
28	19:10	LAST QUARTER MOON
30	05:19	Spica 1.9°N of Moon

Almanac of Astronomical Events for 2068

Date	GMT (h:m)	Event
Jan 04	15	Earth at Perihelion: 0.98331 AU
04	16	Quadrantid Meteor Shower
05	02:38	NEW MOON
06	20:29	Moon at Perigee: 361414 km
08	16	Mercury at Greatest Elong: 19.2°E
11	17:47	FIRST QUARTER MOON
12	07:02	Mars 2.8°S of Moon
14	14:09	Pleiades 3.1°N of Moon
14	22	Jupiter in Conjunction with Sun
19	09:45	FULL MOON
22	15:47	Moon at Apogee: 406042 km
24	15	Mercury at Inferior Conjunction
26	13:08	Spica 1.7°N of Moon
27	14:27	LAST QUARTER MOON
Feb 02	08:33	Mercury 0.5°S of Moon: Occn.
03	13:44	NEW MOON
04	05:46	Moon at Perigee: 357542 km
09	16:04	Mars 1.0°S of Moon: Occn.
10	05:20	FIRST QUARTER MOON
10	19:43	Pleiades 3.4°N of Moon
17	18	Mercury 0.6°N of Jupiter
18	03	Mercury at Greatest Elong: 26.3°W
18	04:38	FULL MOON
18	20:46	Moon at Apogee: 406494 km
22	19:22	Spica 1.4°N of Moon
26	06:25	LAST QUARTER MOON
Mar 01	12:00	Jupiter 4.8°S of Moon
03	18:11	Moon at Perigee: 356802 km
03	23:38	NEW MOON
07	07:22	Venus 0.4°S of Moon: Occn.
08	04:55	Mars 2.7°S of Pleiades
09	03:01	Pleiades 3.6°N of Moon
09	05:13	Mars 0.9°N of Moon: Occn.
10	11	Saturn at Opposition
10	18	Venus at Greatest Elong: 46.3°E
10	19:26	FIRST QUARTER MOON
16	22:03	Moon at Apogee: 406376 km
18	22:56	FULL MOON
19	18:51	Vernal Equinox
21	01:03	Spica 1.3°N of Moon
26	18:20	LAST QUARTER MOON

Date	GMT (h:m)	Event
Apr 01	04:33	Moon at Perigee: 359256 km
02	08:51	NEW MOON
03	04	Mercury at Superior Conjunction
05	11:30	Venus 4.5°N of Moon
05	12:25	Pleiades 3.8°N of Moon
05	21:29	Venus 0.7°N of Pleiades
06	21:38	Mars 2.7°N of Moon
09	11:33	FIRST QUARTER MOON
13	09:52	Moon at Apogee: 405623 km
17	07:21	Spica 1.2°N of Moon
17	15:29	FULL MOON
22	13	Lyrid Meteor Shower
25	02:30	LAST QUARTER MOON
28	22:41	Mercury 1.4°S of Pleiades
29	06:56	Moon at Perigee: 363942 km
29	14	Mercury at Greatest Elong: 20.6°E
May 01	18:07	NEW MOON
02	22:38	Pleiades 3.9°N of Moon
03	05:26	Mercury 2.8°N of Moon
05	03	Eta-Aquarid Meteor Shower
05	16:23	Mars 4.3°N of Moon
09	04:47	FIRST QUARTER MOON
10	20	Mercury 2.7°S of Venus
11	03:18	Moon at Apogee: 404661 km
14	14:48	Spica 1.3°N of Moon
17	05:35	FULL MOON
17	05:40	Partial Lunar Eclipse; mag=0.953
20	20	Venus at Inferior Conjunction
22	02	Mercury at Inferior Conjunction
24	08:00	LAST QUARTER MOON
26	09:16	Moon at Perigee: 368748 km
31	03:54	Total Solar Eclipse; mag=1.011
31	04:03	NEW MOON
Jun 02	12	Uranus at Opposition
07	22:05	Moon at Apogee: 404221 km
07	22:20	FIRST QUARTER MOON
10	23:04	Spica 1.3°N of Moon
15	17:00	FULL MOON
16	21	Mercury at Greatest Elong: 23.2°W
20	10:58	Moon at Perigee: 368627 km
20	11:55	Summer Solstice
22	03:29	Mercury 2.6°N of Aldebaran
22	12:25	LAST QUARTER MOON
26	12:06	Venus 4.2°S of Moon
26	15:08	Pleiades 3.9°N of Moon
27	00	Neptune in Conjunction with Sun
28	00:51	Mercury 0.2°S of Moon: Occn.
29	15:11	NEW MOON

Date	GMT (h:m)	Event
Jul 04	05	Earth at Aphelion: 1.01671 AU
05	16:27	Moon at Apogee: 404637 km
07	15:31	FIRST QUARTER MOON
08	07:19	Spica 1.1°N of Moon
15	02:07	FULL MOON
16	06:09	Venus 1.5°N of Aldebaran
17	00	Mercury at Superior Conjunction
17	14:05	Moon at Perigee: 363871 km
20	23	Mars at Aphelion: 1.66605 AU
21	17:22	LAST QUARTER MOON
23	20:45	Pleiades 4.1°N of Moon
25	07:02	Venus 2.2°S of Moon
28	04	Delta-Aquarid Meteor Shower
29	03:55	NEW MOON
30	00	Venus at Greatest Elong: 45.7°W
31	19:41	Mars 0.6°N of Regulus
Aug 02	08:59	Moon at Apogee: 405607 km
03	09	Jupiter at Opposition
03	19:23	Mercury 0.7°N of Regulus
04	14:46	Spica 0.9°N of Moon
05	13	Mercury 0.1°S of Mars
06	07:38	FIRST QUARTER MOON
12	20	Perseid Meteor Shower
13	09:51	FULL MOON
14	15:33	Moon at Perigee: 359376 km
19	05	Mercury 2.9°S of Saturn
20	00:16	LAST QUARTER MOON
20	02:14	Pleiades 4.3°N of Moon
23	20:56	Venus 1.9°N of Moon
27	18:28	NEW MOON
27	22	Mercury at Greatest Elong: 27.3°E
29	20:38	Moon at Apogee: 406437 km
30	07:10	Mercury 1.4°N of Moon
31	21:10	Spica 0.7°N of Moon
Sep 04	22:04	FIRST QUARTER MOON
11	17:19	FULL MOON
12	00:17	Moon at Perigee: 356950 km
17	05:26	Aldebaran 4.1°S of Moon
18	10:16	LAST QUARTER MOON
19	10	Saturn in Conjunction with Sun
22	04:09	Autumnal Equinox
23	12	Mercury at Inferior Conjunction
26	00:00	Moon at Apogee: 406613 km
26	10:48	NEW MOON
28	03:00	Spica 0.5°N of Moon
30	00:29	Venus 0.0°N of Regulus

Date	GMT (h:m)	Event
Oct 01	22	Mars in Conjunction with Sun
04	10:23	FIRST QUARTER MOON
09	02	Mercury at Greatest Elong: 18.0°W
10	00	Mercury 0.4°S of Saturn
10	11:20	Moon at Perigee: 357423 km
11	01:39	FULL MOON
14	14:14	Aldebaran 4.0°S of Moon
18	00:00	LAST QUARTER MOON
21	13	Orionid Meteor Shower
23	04:48	Moon at Apogee: 406142 km
26	04:17	NEW MOON
26	09	Venus 0.5°S of Saturn
Nov 02	20:38	FIRST QUARTER MOON
05	13	S Taurid Meteor Shower
07	20:44	Moon at Perigee: 360864 km
08	23	Mercury at Superior Conjunction
09	11:40	FULL MOON
09	11:45	Total Lunar Eclipse; mag=1.015
11	00:46	Aldebaran 3.9°S of Moon
12	12	N Taurid Meteor Shower
13	12:27	Venus 3.5°N of Spica
16	17:33	LAST QUARTER MOON
17	19	Leonid Meteor Shower
19	19:10	Moon at Apogee: 405230 km
21	16:10	Spica 0.5°N of Moon
22	16:46	Venus 3.4°N of Moon
23	09:34	Mars 1.1°N of Moon: Occn.
24	21:30	Partial Solar Eclipse; mag=0.911
24	21:42	NEW MOON
Dec 02	05:21	FIRST QUARTER MOON
05	19:40	Moon at Perigee: 366315 km
06	08	Uranus in Conjunction with Sun
08	10	Venus 1.1°N of Mars
08	11:11	Aldebaran 3.9°S of Moon
08	23:42	FULL MOON
14	08	Geminid Meteor Shower
16	14:11	LAST QUARTER MOON
17	15:04	Moon at Apogee: 404457 km
19	00:05	Spica 0.5°N of Moon
21	01:34	Winter Solstice
21	21	Mercury at Greatest Elong: 20.1°E
22	07:21	Mars 1.0°S of Moon: Occn.
22	16	Ursid Meteor Shower
22	23:48	Venus 1.0°S of Moon: Occn.
24	13:44	NEW MOON
28	07	Neptune at Opposition
31	13:23	FIRST QUARTER MOON
31	15:07	Moon at Perigee: 370337 km

Almanac of Astronomical Events for 2069

Date	GMT (h:m)	Event
Jan 03	22	Quadrantid Meteor Shower
04	19	Earth at Perihelion: 0.98329 AU
04	19:33	Aldebaran 3.9°S of Moon
07	13:43	FULL MOON
07	18	Mercury at Inferior Conjunction
14	12:33	Moon at Apogee: 404393 km
15	08:13	Spica 0.2°N of Moon
15	12:16	LAST QUARTER MOON
20	07:10	Mars 3.1°S of Moon
21	10:03	Mercury 1.3°S of Moon
23	03:36	NEW MOON
26	07:47	Moon at Perigee: 366478 km
29	21:39	FIRST QUARTER MOON
30	11	Mercury at Greatest Elong: 25.0°W
Feb 01	01:32	Aldebaran 3.7°S of Moon
06	05:29	FULL MOON
11	08:23	Moon at Apogee: 405051 km
11	15:53	Spica 0.0°S of Moon
14	09:27	LAST QUARTER MOON
17	10	Jupiter in Conjunction with Sun
21	15:17	NEW MOON
23	06:27	Moon at Perigee: 361058 km
28	06:54	FIRST QUARTER MOON
28	06:55	Aldebaran 3.4°S of Moon
Mar 07	22:35	FULL MOON
10	22:33	Moon at Apogee: 405880 km
10	22:44	Spica 0.2°S of Moon
11	12	Venus at Superior Conjunction
16	03:31	LAST QUARTER MOON
17	16	Mercury at Superior Conjunction
20	00:44	Vernal Equinox
23	01:13	NEW MOON
23	15	Saturn at Opposition
23	15:45	Moon at Perigee: 357672 km
27	14:00	Aldebaran 3.2°S of Moon
29	17:34	FIRST QUARTER MOON

Date	GMT (h:m)	Event
Apr 06	16:13	FULL MOON
07	02:44	Moon at Apogee: 406308 km
07	04:59	Spica 0.3°S of Moon
12	04	Mercury at Greatest Elong: 19.5°E
14	17:21	LAST QUARTER MOON
21	02:57	Moon at Perigee: 357266 km
21	09:58	NEW MOON
21	10:09	Partial Solar Eclipse; mag=0.899
22	20	Lyrid Meteor Shower
23	23:29	Aldebaran 3.1°S of Moon
28	05:56	FIRST QUARTER MOON
May 02	06	Mercury at Inferior Conjunction
04	05:38	Moon at Apogee: 406134 km
04	11:08	Spica 0.3°S of Moon
05	09	Eta-Aquarid Meteor Shower
06	09:08	Total Lunar Eclipse; mag=1.323
06	09:11	FULL MOON
11	20	Mars 0.7°S of Jupiter
14	03:10	LAST QUARTER MOON
19	07:36	Mercury 3.9°S of Moon
19	12:00	Moon at Perigee: 359730 km
20	17:51	Partial Solar Eclipse; mag=0.088
20	18:06	NEW MOON
22	01:32	Venus 3.8°N of Moon
27	20:09	FIRST QUARTER MOON
29	11	Mercury at Greatest Elong: 24.9°W
31	17:30	Moon at Apogee: 405336 km
31	17:43	Spica 0.3°S of Moon
Jun 05	00:19	FULL MOON
07	09	Uranus at Opposition
12	09:56	LAST QUARTER MOON
13	22:42	Mars 4.9°S of Moon
16	13:22	Moon at Perigee: 364199 km
17	20:05	Aldebaran 3.1°S of Moon
17	22:48	Mercury 1.5°N of Moon
19	02:14	NEW MOON
20	17:40	Summer Solstice
26	12:10	FIRST QUARTER MOON
28	00:56	Spica 0.4°S of Moon
28	10:15	Moon at Apogee: 404412 km
29	09	Mars at Perihelion: 1.38120 AU
29	14	Neptune in Conjunction with Sun

Date	GMT (h:m)	Event
Jul 01	09	Mercury at Superior Conjunction
04	13:05	FULL MOON
05	23	Earth at Aphelion: 1.01665 AU
11	14:59	LAST QUARTER MOON
12	13:02	Mars 2.9°S of Moon
13	15:08	Moon at Perigee: 368840 km
15	04:02	Aldebaran 3.0°S of Moon
18	11:13	NEW MOON
18	21:20	Venus 1.0°N of Regulus
25	08:38	Spica 0.7°S of Moon
26	04:43	Moon at Apogee: 404051 km
26	05:30	FIRST QUARTER MOON
28	06:07	Mercury 0.0°N of Regulus
28	10	Delta-Aquarid Meteor Shower
Aug 02	23:44	FULL MOON
07	17:00	Moon at Perigee: 368723 km
09	19:41	LAST QUARTER MOON
10	00:15	Mars 0.8°S of Moon: Occn.
10	08	Mercury at Greatest Elong: 27.4°E
11	09:58	Aldebaran 2.8°S of Moon
13	02	Perseid Meteor Shower
16	22:03	NEW MOON
17	03	Venus 1.8°S of Saturn
19	01:49	Mercury 0.3°N of Moon: Occn.
20	11:54	Venus 2.8°N of Moon
21	16:21	Spica 0.9°S of Moon
22	23:32	Moon at Apogee: 404552 km
24	23:17	FIRST QUARTER MOON
Sep 01	09:06	FULL MOON
03	04:39	Venus 1.2°N of Spica
03	20:15	Moon at Perigee: 363898 km
06	17	Mercury at Inferior Conjunction
07	07:19	Mars 1.0°N of Moon: Occn.
07	15:19	Aldebaran 2.6°S of Moon
08	01:22	LAST QUARTER MOON
09	20	Jupiter at Opposition
15	11:35	NEW MOON
17	23:36	Spica 1.1°S of Moon
19	11:56	Venus 2.6°S of Moon
19	17:06	Moon at Apogee: 405575 km
22	09:51	Autumnal Equinox
22	18	Mercury at Greatest Elong: 17.9°W
23	16:23	FIRST QUARTER MOON
30	18:09	FULL MOON

Date	GMT (h:m)	Event
Oct 01	22:56	Moon at Perigee: 359277 km
02	15	Saturn in Conjunction with Sun
04	22:08	Aldebaran 2.4°S of Moon
05	07:12	Mars 2.6°N of Moon
07	09:20	LAST QUARTER MOON
15	04:03	NEW MOON
15	04:18	Partial Solar Eclipse; mag=0.530
15	08	Venus at Greatest Elong: 46.8°E
15	20:35	Venus 0.9°N of Antares
17	05:28	Moon at Apogee: 406378 km
20	12	Mercury at Superior Conjunction
21	19	Orionid Meteor Shower
23	07:57	FIRST QUARTER MOON
30	03:33	Total Lunar Eclipse; mag=1.462
30	03:35	FULL MOON
30	09:15	Moon at Perigee: 356831 km
Nov 01	07:35	Aldebaran 2.4°S of Moon
01	18:40	Mars 3.9°N of Moon
05	19	S Taurid Meteor Shower
05	20:40	LAST QUARTER MOON
10	21:25	Saturn 4.3°N of Moon
11	12:13	Spica 1.1°S of Moon
12	19	N Taurid Meteor Shower
13	07:36	Moon at Apogee: 406514 km
13	22:38	NEW MOON
16	09:54	Mercury 2.5°N of Antares
18	01	Leonid Meteor Shower
21	21:31	FIRST QUARTER MOON
27	21:53	Moon at Perigee: 357487 km
28	13:46	FULL MOON
28	16:36	Mars 4.6°N of Moon
28	18:53	Aldebaran 2.4°S of Moon
30	08	Mars at Opposition
Dec 04	20	Mercury at Greatest Elong: 21.2°E
05	12:03	LAST QUARTER MOON
08	08:47	Saturn 4.0°N of Moon
08	18:25	Spica 1.2°S of Moon
10	12:50	Moon at Apogee: 406042 km
10	21	Uranus in Conjunction with Sun
13	17:38	NEW MOON
14	14	Geminid Meteor Shower
21	07:21	Winter Solstice
21	09:00	FIRST QUARTER MOON
22	23	Ursid Meteor Shower
23	00	Mercury at Inferior Conjunction
25	02	Venus at Inferior Conjunction
25	14:36	Mars 4.9°N of Moon
26	05:46	Aldebaran 2.4°S of Moon
26	07:40	Moon at Perigee: 361240 km
28	00:50	FULL MOON
30	17	Neptune at Opposition

Almanac of Astronomical Events for 2070

Date	GMT (h:m)	Event
Jan 03	03	Earth at Perihelion: 0.98334 AU
04	04	Quadrantid Meteor Shower
04	07:16	LAST QUARTER MOON
04	16:55	Mars 1.3°S of Pleiades
04	19:13	Saturn 3.7°N of Moon
05	01:26	Spica 1.4°S of Moon
07	04:45	Moon at Apogee: 405122 km
09	00	Mercury 3.3°S of Venus
10	09:14	Venus 1.4°N of Moon
10	10:19	Mercury 2.2°S of Moon
12	11:22	NEW MOON
12	20	Mercury at Greatest Elong: 23.5°W
17	00:39	Jupiter 4.6°S of Moon
19	18:31	FIRST QUARTER MOON
22	14:12	Aldebaran 2.2°S of Moon
23	04:43	Moon at Perigee: 366854 km
26	12:59	FULL MOON
Feb 01	04:19	Saturn 3.5°N of Moon
01	09:31	Spica 1.7°S of Moon
03	04:46	LAST QUARTER MOON
04	01:19	Moon at Apogee: 404362 km
07	11:23	Venus 0.5°N of Moon: Occn.
11	02:52	NEW MOON
13	16:37	Jupiter 3.9°S of Moon
17	15:55	Moon at Perigee: 370239 km
18	02:33	FIRST QUARTER MOON
18	20:05	Aldebaran 2.0°S of Moon
25	02:31	FULL MOON
28	11:24	Saturn 3.5°N of Moon
28	14	Mercury at Superior Conjunction
28	18:02	Spica 1.9°S of Moon
Mar 03	22:21	Moon at Apogee: 404337 km
05	02:11	LAST QUARTER MOON
05	17	Venus at Greatest Elong: 46.7°W
08	23:40	Venus 2.8°S of Moon
12	15:52	NEW MOON
15	17:55	Moon at Perigee: 366070 km
18	01:32	Aldebaran 1.8°S of Moon
19	09:53	FIRST QUARTER MOON
20	06:35	Vernal Equinox
26	01	Jupiter in Conjunction with Sun
26	04	Mercury at Greatest Elong: 18.7°E
26	17:31	FULL MOON
27	16:03	Saturn 3.7°N of Moon
28	01:59	Spica 2.0°S of Moon
31	17:01	Moon at Apogee: 405036 km

Date	GMT (h:m)	Event
Apr 03	21:23	LAST QUARTER MOON
05	14	Saturn at Opposition
07	21:51	Venus 4.5°S of Moon
11	02:30	NEW MOON
11	02:34	Total Solar Eclipse; mag=1.047
12	16:43	Moon at Perigee: 360956 km
13	05	Mercury at Inferior Conjunction
14	08:51	Aldebaran 1.8°S of Moon
17	17:32	FIRST QUARTER MOON
23	02	Lyrid Meteor Shower
23	18:52	Saturn 3.9°N of Moon
24	08:45	Spica 2.0°S of Moon
25	09:19	Pen. Lunar Eclipse; mag=1.052
25	09:31	FULL MOON
28	06:11	Moon at Apogee: 405893 km
29	02	Mercury 3.5°N of Jupiter
May 03	13:11	LAST QUARTER MOON
05	15	Eta-Aquarid Meteor Shower
07	18:17	Venus 3.3°S of Moon
08	05:24	Jupiter 2.1°S of Moon
08	16:44	Mercury 3.5°S of Moon
10	11:08	NEW MOON
11	01:00	Moon at Perigee: 357790 km
11	03	Mercury at Greatest Elong: 26.3°W
11	18:33	Aldebaran 1.8°S of Moon
15	05	Venus 0.7°S of Jupiter
17	02:30	FIRST QUARTER MOON
20	21:38	Saturn 3.9°N of Moon
21	14:36	Spica 2.0°S of Moon
25	01:37	FULL MOON
25	10:55	Moon at Apogee: 406329 km
Jun 02	01:26	LAST QUARTER MOON
05	00:22	Jupiter 1.5°S of Moon
06	11:53	Venus 0.1°S of Moon: Occn.
07	19	Mars at Aphelion: 1.66601 AU
08	10:48	Moon at Perigee: 357376 km
08	18:24	NEW MOON
12	05	Uranus at Opposition
15	13:40	FIRST QUARTER MOON
15	21	Mercury at Superior Conjunction
17	02:32	Saturn 3.7°N of Moon
17	20:27	Spica 2.2°S of Moon
20	23:22	Summer Solstice
21	14:15	Moon at Apogee: 406125 km
23	16:57	FULL MOON

Date	GMT (h:m)	Event
Jul 01	10:33	LAST QUARTER MOON
02	04	Neptune in Conjunction with Sun
02	15:40	Jupiter 0.9°S of Moon: Occn.
05	15:28	Aldebaran 1.7°S of Moon
06	06:19	Venus 3.3°N of Moon
06	12	Earth at Aphelion: 1.01671 AU
06	18:57	Moon at Perigee: 359714 km
08	01:14	NEW MOON
11	21	Jupiter at Perihelion: 4.94826 AU
12	16:01	Mars 0.6°N of Regulus
14	10:50	Saturn 3.3°N of Moon
15	03:14	Spica 2.4°S of Moon
15	03:26	FIRST QUARTER MOON
19	01:07	Moon at Apogee: 405342 km
23	07:02	FULL MOON
23	13	Mercury at Greatest Elong: 27.0°E
25	02:33	Mercury 1.5°S of Regulus
28	17	Delta-Aquarid Meteor Shower
30	02:20	Jupiter 0.5°S of Moon: Occn.
30	17:17	LAST QUARTER MOON
Aug 01	23:30	Aldebaran 1.5°S of Moon
03	20:19	Moon at Perigee: 364129 km
06	08:51	NEW MOON
07	19:46	Mercury 0.6°S of Moon: Occn.
08	18:53	Mars 3.7°N of Moon
10	22:29	Saturn 2.7°N of Moon
11	11:16	Spica 2.7°S of Moon
13	08	Perseid Meteor Shower
13	19:40	FIRST QUARTER MOON
15	17:17	Moon at Apogee: 404491 km
20	08	Mercury at Inferior Conjunction
21	19:54	FULL MOON
26	08:36	Jupiter 0.4°S of Moon: Occn.
28	22:41	LAST QUARTER MOON
29	05:27	Aldebaran 1.3°S of Moon
30	22:39	Moon at Perigee: 368889 km
Sep 03	11:52	Mercury 3.6°N of Moon
04	18:29	NEW MOON
06	08	Mercury at Greatest Elong: 18.0°W
06	11:56	Mars 1.8°N of Moon
07	12:13	Saturn 2.3°N of Moon
07	20:02	Spica 2.8°S of Moon
09	13:19	Mercury 0.3°N of Regulus
12	12:02	Moon at Apogee: 404226 km
12	13:44	FIRST QUARTER MOON
20	07:47	FULL MOON
22	12:11	Jupiter 0.5°S of Moon: Occn.
22	15:45	Autumnal Equinox
24	22:02	Moon at Perigee: 368697 km
25	10:51	Aldebaran 1.2°S of Moon
27	04:02	LAST QUARTER MOON

Date	GMT (h:m)	Event
Oct 02	03	Mercury at Superior Conjunction
04	07:01	NEW MOON
04	07:07	Annular Solar Eclipse; mag=0.973
08	12	Venus at Superior Conjunction
10	07:45	Moon at Apogee: 404811 km
12	08:40	FIRST QUARTER MOON
15	10	Saturn in Conjunction with Sun
17	15	Jupiter at Opposition
19	15:42	Jupiter 0.9°S of Moon: Occn.
19	18:49	Partial Lunar Eclipse; mag=0.138
19	18:59	FULL MOON
22	01	Orionid Meteor Shower
22	03:04	Moon at Perigee: 363541 km
22	17:50	Aldebaran 1.2°S of Moon
26	10:47	LAST QUARTER MOON
Nov 01	11:37	Spica 2.8°S of Moon
01	15:46	Saturn 1.6°N of Moon
02	22:42	NEW MOON
06	01	S Taurid Meteor Shower
07	02:03	Moon at Apogee: 405849 km
10	14:04	Mercury 2.0°N of Antares
10	17	Mars in Conjunction with Sun
11	03:20	FIRST QUARTER MOON
13	01	N Taurid Meteor Shower
15	21:06	Jupiter 1.1°S of Moon: Occn.
17	14	Mercury at Greatest Elong: 22.5°E
18	05:40	FULL MOON
18	07	Leonid Meteor Shower
19	03:31	Aldebaran 1.3°S of Moon
19	07:55	Moon at Perigee: 358729 km
24	17:36	Regulus 4.3°N of Moon
24	20:20	LAST QUARTER MOON
28	17:27	Spica 2.9°S of Moon
29	03:29	Saturn 1.3°N of Moon
Dec 02	16:53	NEW MOON
04	13:36	Moon at Apogee: 406585 km
07	07	Mercury at Inferior Conjunction
10	20:32	FIRST QUARTER MOON
13	04:49	Jupiter 0.9°S of Moon: Occn.
14	20	Geminid Meteor Shower
15	10	Uranus in Conjunction with Sun
16	14:55	Aldebaran 1.3°S of Moon
17	16:05	FULL MOON
17	19:41	Moon at Perigee: 356442 km
21	13:19	Winter Solstice
22	01:05	Regulus 4.1°N of Moon
23	05	Ursid Meteor Shower
24	09:31	LAST QUARTER MOON
25	23:03	Spica 3.1°S of Moon
26	08	Mercury at Greatest Elong: 22.1°W
26	13:43	Saturn 0.9°N of Moon: Occn.
30	11:50	Mercury 3.1°S of Moon
31	14:10	Moon at Apogee: 406681 km

Accuracy of the Almanac of Astronomical Events

The goal of the Almanac of Astronomical Events is to present a wide range of solar system phenomena with practicable accuracy. In general, events listed to the nearest hour are accurate to ± 30 minutes. Events listed in hours and minutes (i.e., hh:mm) are typically accurate to ± 5 minutes or less.

The following table gives a detailed breakdown of the time accuracy of various astronomical events.

Astronomical Event	Accuracy
Solar and Lunar Eclipses	± 0.5 minute
Phases of the Moon	± 0.5 minute
Apogee/Perigee of Moon	± 5 minutes; ± 5 kilometers
Solstice/Equinox (Earth)	± 0.5 minute
Aphelion/Perihelion (Earth)	± 30 minutes; ± 0.00001 AU
Conjunctions of Moon with Star or Planet	± 10 minutes
Conjunctions of Planet with Planet	± 3 hours
Inferior/Superior Conjunctions (Mercury & Venus)	± 30 minutes
Greatest Elongation (Mercury & Venus)	± 30 minutes
Opposition/Conjunction (Outer Planets)	± 3 hours
Aphelion/Perihelion of Planets	± 30 minutes

Algorithms used in predicting many of the astronomical events are based on Astronomical Algorithms by Jean Meeus (Willmann-Bell Inc. Richmond 1998). When needed, use has been made of the JPL DE405 (e.g., solar and lunar eclipses, planetary aphelion and perihelion).

AstroPixels.com and the Sky Event Almanacs

The Astropixels Web Site was created as a place to post my astrophotography from Arizona. But I soon expanded it to include handy resources for solar system objects. Tables for Earth's equinoxes and solstices, phases of the Moon, apogees and perigees of the Moon, and ephemerides for the Sun, Moon and planets are some of the pages found there.

Most recently, I added the *Sky Event Almanacs* - a collection of 2400 web pages. Each page lists over two hundred major astronomical events for a single year and for a single time zone. The *Sky Event Almanacs* cover every year of the 21st Century and for 24 time zones around the world. Just select a time zone and a year and see a concise and convenient list of sky events for that period.

The astronomical phenomena listed include the following.

- solar and lunar eclipses
- phases of the Moon
- apogees and perigees of the Moon
- Equinoxes and Solstices of Earth
- aphelion and perihelion (Earth, Mars, Jupiter, Saturn and Uranus)
- oppositions and conjunctions of the planets
- elongations of Mercury and Venus
- close conjunctions of the Moon with the planets and bright stars
- close conjunctions of planets with bright stars and other planets
- peak of major meteor showers

FIFTY YEAR ALMANAC OF ASTRONOMICAL EVENTS — 2021 TO 2070

All of these events are conveniently given for the time zone and year of choice. The main directory page of the *Sky Event Almanac* appears below. This page has links to all 24 time zones and all 100 years.

http://www.astropixels.com/almanac/almanac.html

The *Fifty Year Almanac of Astronomical Events – 2021 To 2070* is a direct spin-off from the *Sky Event Almanac*. It is convenient reference covering five decades and for Greenwich Mean Time.

Below are a few useful links to *Astropixels.com*:

Astropixels Home Page: *http://www.astropixels.com/index.html*
Sky Event Almanacs Index: *http://www.astropixels.com/almanac/almanac.html*
Planetary Ephemeris Data: *http://www.astropixels.com/ephemeris/ephemeris.html*
Observing Resources Page: *http://www. astropixels.com/main/resources.html*
Astropixels Photo Index: *http://www. astropixels.com/main/photoindex.html*
Astropixels Publishing Index: *http://www. astropixels.com/pubs/index.html*

EclipseWise.com Web Site

I created and maintained the NASA Eclipse Web Site for many years. After I retired, I decided to design a new website that would go far beyond the old NASA site both in scope and features. In 2014, I launched the new eclipse prediction website and named it *EclipseWise.com*.

EclipseWise.com has individual web pages, maps and diagrams for every solar and lunar eclipse from 2000 BCE to 3000 CE. This covers to 11,898 solar eclipses and 12,064 lunar eclipses. Much of the design, layout and graphics were inspired by the recent publications *Thousand Year Canon of Solar Eclipses 1501 to 2500* and the *Thousand Year Canon of Lunar Eclipses 1501 to 2500*.

The graphical user interface used by EclipseWise.com offers an intuitive way of accessing eclipse predictions. For example, the home page presents a concise preview of all upcoming solar and lunar eclipses over several years. Each small eclipse diagram gives a quick preview of an eclipse and links to a dedicated page for that particular eclipse.

The main or top pages of *EclipseWise.com* are:

Home Page (both solar and lunar eclipses):
    http://www.eclipsewise.com/eclipse.html
Solar Eclipses Page:
    http://www.eclipsewise.com/solar/solar.html
Lunar Eclipses Page:
    http://www.eclipsewise.com/lunar/lunar.html

Detailed information on solar and lunar eclipse photography, and tips on eclipse observing and eye safety may be found at my *MrEclipse* website:

http://www.MrEclipse.com/MrEclipse.html

Many images of recent solar and lunar eclipses are posted on *MrEclipse.com*.

Astropixels Publications

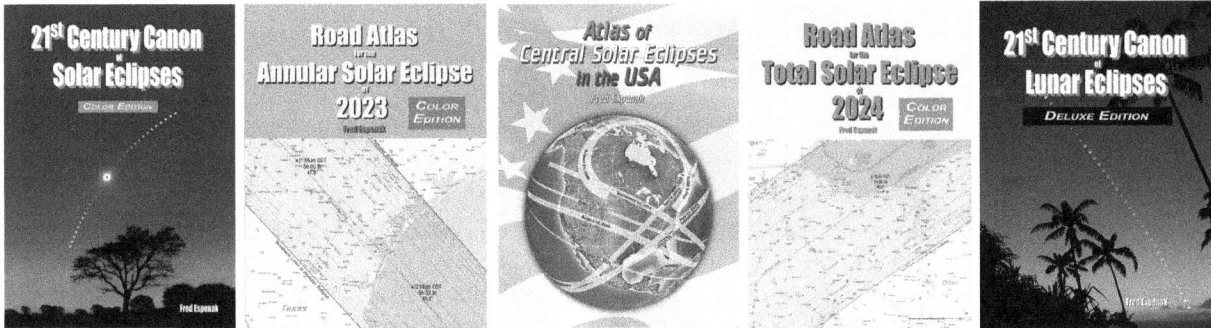

Below is a partial list of some of the books Fred Espenak has written through Astropixels Publishing:

21st Century Canon of Solar Eclipses

The complete guide to every solar eclipse occurring from 2001 tom 2100 (224 eclipses in all). It includes information and maps for all total, annular, hybrid, and partial eclipses. A special world atlas shows detailed full page maps of all central eclipse paths (total, annular and hybrid).

Road Atlas for the Annular Solar Eclipse of 2023

Detailed road maps of the entire eclipse path from the western USA, through Mexico, Central and South America. Information printed on the maps makes it easy to estimate the duration of annularity from any location in the eclipse path. This is the next annular solar eclipse visible from the USA.

Atlas of Central Solar Eclipses in the USA

When was the last total eclipse through the USA and when is the next? How often do they happen? What total eclipse tracks passed across the USA during the 17th, 18th, and 19th centuries, etc., and what states did they include? And how often is a total solar eclipse visible from each of the 50 states? The Atlas of Central Solar Eclipses in the USA answers all of these questions and more with hundreds of maps and tables.

Road Atlas for the Total Solar Eclipse of 2024

This book contains detailed road maps of the entire eclipse path from Mexico, through the USA and Canada. Information printed on the maps makes it easy to estimate the duration of totality from any location in the eclipse path. This is the next total solar eclipse visible from the USA.

21st Century Canon of Lunar Eclipses

The complete guide to every lunar eclipse occurring from 2001 tom 2100 (228 eclipses in all). It includes information and maps for all total, partial, and penumbral eclipses. The predictions use a new model for Earth's elliptical shadows.

For information on these books and more, visit *Astropixels Publishing*:

http://eclipsewise.com/pubs/index.html

All Astropixels Publications are available from Amazon.com

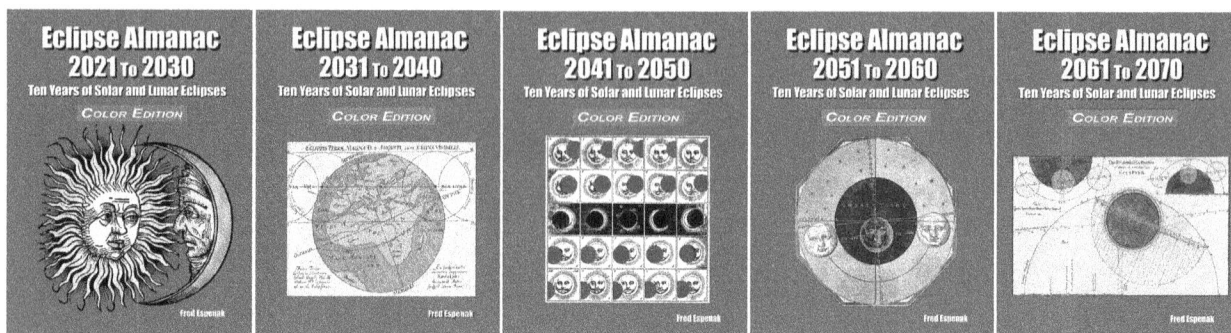

The **Eclipse Almanac** contains maps and diagrams of every solar and lunar eclipse over a ten-year period. This permits the reader to look ahead and easily determine when and where each of these events will be seen. Particular details about each eclipse are included as well as a 25-year table looking further into the future.

Section 1 covers solar eclipses, while Section 2 is devoted to lunar eclipses. Explanations are given for each type of eclipse, and the visual appearance of each one is included. Section 3 tabulates the date and time of the Moon's phases over a decade. New Moon and Full Moon phases that coincide with solar and lunar eclipses, are identified.

The **Eclipse Almanac** series consists of five volumes, each available in a color, black and white, and Kindle editions. Each volume covers a single decade:

1) Eclipse Almanac 2021 to 2030	4) Eclipse Almanac 2051 to 2060
2) Eclipse Almanac 2031 to 2040	5) Eclipse Almanac 2061 to 2070
3) Eclipse Almanac 2041 to 2050	

All Astropixels Publications are available from Amazon.com

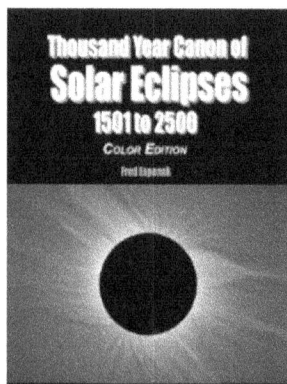

The **Thousand Year Canon of Solar Eclipses 1501 to 2500** contains maps and data for each of the 2,389 solar eclipses occurring over the ten century period centered on the present era.

The primary content of the Thousand Year Canon is a comprehensive catalog listing the essential characteristics of each eclipse and an atlas of 2,389 maps depicting the geographic regions of visibility of each eclipse.

Available in either black & white or color editions.

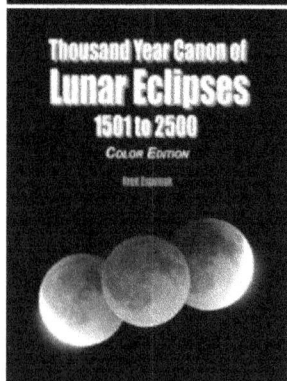

The **Thousand Year Canon of Lunar Eclipses 1501 to 2500** contains diagrams, maps and data for each of the 2,424 lunar eclipses occurring over the ten century period centered on the present era.

Appendix A is a comprehensive catalog listing the essential characteristics of each eclipse. Appendix B is an atlas depicting the path of the Moon through Earth's shadows and maps identifying the geographic regions of visibility of each of the 2,424 eclipses.

Available in either black & white or color editions.

All Astropixels Publications are available from Amazon.com